ESSENTIAL
MATH SKILLS

A MATHEMATICS COMPETENCY WORKBOOK
THIRD EDITION

LEO GAFNEY • JOHN C. BEERS

Phoenix Learning Resources
New York • St. Louis

Design, Production, and Editorial Services: Pencil Point Studio

ISBN: 0-7915-3198-8

 2 3 4 5 6 7 8 9 0 05 04 03 02 01

CONTENTS

Chapter 6   GRAPHS AND FORMULAS

Chapter 7   GEOMETRY

Chapter 8   RATIONAL NUMBERS AND EQUATIONS

PRETEST

Add.

1. 38	**2.** 71	**3.** 9	**4.** 28	**5.** 185	**6.** 61	**7.** 26 + 176 = _____
+51	+ 0	+7	+33	+349	93	
					+28	

Subtract.

8. 65	**9.** 43	**10.** 80	**11.** 302	**12.** 426	**13.** 93 − (52 − 1) = _____
−10	−17	−26	−185	−128	

Multiply.

14. 11	**15.** 24	**16.** 54	**17.** 326	**18.** 627	**19.** 23 x 45 = _____
x 4	x 3	x 7	x 53	x 90	

Divide.   **Simplify.**

20. $2\overline{)46}$ **21.** $7\overline{)714}$ **24.** 3 + 4 x 5   **25.** 6(7-2) + 1

22. $18\overline{)594}$ **23.** $23\overline{)617}$ **26.** Evaluate 2(a + b) for a=2, b=5

Raise to the power shown.

27. 6^2 = _____   **28.** 5^3 = _____

Round to the nearest hundred.

29. 341 _____   **30.** 587 _____   **31.** 1550 _____

Write the numbers.

32. Six thousand twenty _____   **33.** Twenty thousand two hundred twelve _____

ADDING WHOLE NUMBERS

907	352
+ 61	+634
968	986

Add units first; then tens; then hundreds.

Add.

1. 32 +54	**2.** 28 +70	**3.** 43 +25	**4.** 26 +61	**5.** 38 +20	**6.** 61 +14
7. 432 +523	**8.** 221 +706	**9.** 563 +216	**10.** 224 +603	**11.** 815 +133	**12.** 629 +120
13. 338 + 60	**14.** 8152 + 743	**15.** 252 + 47	**16.** 7123 + 405	**17.** 1813 + 64	**18.** 528 + 70
19. 4267 +5112	**20.** 7532 +1436	**21.** 2607 +5181	**22.** 6930 +1036	**23.** 4812 +1065	**24.** 8116 +1753

25. Last year, the Madison School Library contained 7843 books. This year 152 new books were purchased. How many books does the library contain now?

_____

27. Ivan drove 2561 miles in April and 1137 in May. How many miles did he drive during the two months?

_____

29. In 1998 the population of Canaan was 7030. Since then the population has increased by 759. What is the population now?

_____

26. The Brightstar Shoe Store had 521 pairs of shoes in the stock room. During an inventory check, 223 were counted on the shelves. 14 more pairs were on display in the window. How many pairs of shoes did the Brightstar Shoe Store have in all?

_____

28. The Corner Store had 305 cases of soup on hand when 72 cases were delivered. How many cases does the store now have?

_____

30. The Card Store sold 4231 baseball cards last month and 2037 this month. How many cards were sold in both months combined?

_____

ADDING WHOLE NUMBERS

397
+625
1022

Whenever the sum of a column is a two–digit number, you have to carry.

Add.

1. 75
    +98

2. 34
    +29

3. 68
    +13

4. 77
    +47

5. 846
    + 93

6. 68
    +896

7. 864
    +219

8. 718
    +198

9. 957
    + 483

10. 6596
    + 481

11. 9309
    + 792

12. 28
    +1988

13. 6475
    +9825

14. 4728
    +7188

15. 7405
    +6913

16. 98,163
    + 4,192

17. 5,591
    +26,259

18. 89,842
    + 6,457

19. 38
    49
    +75

20. 76
    19
    +67

21. 196
    87
    + 123

22. 309
    80
    +911

23. 3864
    619
    +4719

24. 5607
    9216
    + 422

25. The Farmer's market sold 875 pounds of potatoes on Monday, 723 on Tuesday, and 567 on Wednesday. How many pounds were sold on the three days together?

_____

26. The Red Barons hit 187 home runs last year and 209 home runs this year. How many home runs were hit in both years together?

_____

27. Simsbury is 569 miles east of Gainsville. Fairview is 277 miles east of Simsbury. How far is it from Gainsville to Simsbury?

_____

28. The populations of three cities are: Simsbury has 24,308; Fairview has 19,436; Gainsville has 6,289. What is the total population of the three cities?

_____

29. The Simsbury Public Library has 40,728 books. Fairview has 35,206; Gainsville has 40,619. Which town's library has the most books?

_____

30. What is the total number of books in the libraries of the three towns?

_____

SUBTRACTING WHOLE NUMBERS

Subtract.

1. 39 − 7	**2.** 28 − 5	**3.** 26 − 4	**4.** 57 − 4	**5.** 48 −32	**6.** 69 −15
7. 561 − 30	**8.** 197 − 71	**9.** 385 − 43	**10.** 761 − 31	**11.** 879 −533	**12.** 438 −313
13. 7645 − 224	**14.** 4857 − 204	**15.** 8793 − 670	**16.** 9740 −6320	**17.** 1986 −1471	**18.** 7885 −4804
19. 89,654 −23,142	**20.** 81,457 −60,317	**21.** 39,628 −22,312	**22.** 65,948 −41,726	**23.** 84,196 −73,042	**24.** 47,478 −25,158
25. 917,684 − 13,602	**26.** 752,873 − 40,351	**27.** 959,928 −431,805	**28.** 716,885 −603,271	**29.** 991,283 −371,062	**30.** 417,983 −311,780

31. Mary Lou's car weighs 2997 lbs. Jason's car weighs 1752 lbs. How much more does Mary Lou's car weigh than Jason's?

32. The Moving Company van traveled 587 miles on Monday and 444 on Tuesday. How much farther did it travel on Monday?

33. The Earthman Excavation Co. dug out 5628 tons of dirt on Tuesday and 3210 tons on Wednesday. How much more earth did the company dig on Tuesday?

34. Jason wrote a novel with 48,765 words and two years later a novel with 38,705 words. How many more words were in the first novel than the second?

35. The tallest building in Bigtown is 987 ft high. The tallest building in Smalltown is 426 ft. How much taller is the tallest building in Bigtown?

36. The Newton High basketball team scored 1528 points last season and 1739 this season. How many more points were scored last season than this season?

SUBTRACTING WHOLE NUMBERS

Subtract.

1. 85 −48	**2.** 93 −25	**3.** 72 −37	**4.** 47 −39	**5.** 66 −37	**6.** 264 − 38
7. 636 − 72	**8.** 128 − 73	**9.** 237 − 72	**10.** 342 − 37	**11.** 129 − 47	**12.** 275 − 38
13. 631 −294	**14.** 416 −189	**15.** 744 −376	**16.** 823 −346	**17.** 3121 − 647	**18.** 1157 − 289
19. 3625 − 646	**20.** 4251 − 883	**21.** 5254 −2798	**22.** 4346 −3548	**23.** 6526 −5759	**24.** 8535 −3978
25. 25,331 − 8,273	**26.** 24,522 − 7,837	**27.** 25,433 − 6,847	**28.** 22,415 −13,827	**29.** 16,532 − 6,568	**30.** 63,283 − 5,619

31. The Sears Tower in Chicago is 1454 ft tall. The World Trade Center tower in New York is 1368 ft tall. How much taller is the Sears Tower?

32. Alaska has an area of 656,424 sq mi. Texas has an area of 266,807 sq mi. How much greater is the area of Alaska than the area of Texas?

33. The highest mountain in Alaska is 20,320 ft high. Mt. Whitney in California is 14,494 ft high. How much higher is the highest mountain in Alaska than Mt. Whitney?

34. In 1850 the population of New York City was 696,115. In 1990 the population was 7,322,564. By how much did the population increase in those 140 years?

35. Australia has an area of 2,966,200 sq mi. The United States has an area of 3,787,319 sq mi. Which country is larger?

36. Mt. Everest is 29,018 ft high. Mt. Godwin Austen, the second highest mountain in the world, is 28,250 ft high. How much higher is Mt. Everest?

SUBTRACTING WHOLE NUMBERS

Place the larger number on top before subtracting.

EXAMPLES **Subtract 18 from 26.**

$$
\begin{array}{r}
\text{1 16} \\
\cancel{26} \\
-\,18 \\
\hline
8
\end{array}
$$

Subtract.

1. From 200 subtract 152. _____

2. From 237 subtract 74. _____

3. From 9153 take 2847. _____

4. From 37,450 take 15,958. _____

5. Take 63 from 325. _____

6. From 2000 take 405. _____

7. Subtract 2,059 from 37,235. _____

8. Take 69 from 400. _____

9. From 9514 take 7352. _____

10. Subtract 306 from 1583. _____

11. Take 369 from 492. _____

12. Decrease 650 by 189. _____

13. Take 600 from 1647. _____

14. Decrease 8654 by 8195. _____

15. Find the difference between 69,224 and 41,400. _____

16. Find the difference between 69,224 and 41,400. _____

17. The difference between two numbers is 55. One of the numbers is 203. The other number is either

_____ or _____

18. Two numbers together total 13,509. One of the numbers is 6,987. Find the other number.

_____

19. Find a number that is 67 less than 290.

_____

20. Find a number that is 178 more than 593.

_____

21. In 1960, John F. Kennedy won the presidential election over Richard M. Nixon. Kennedy had 34,226,731 votes. Nixon had 34,108,157. How many more votes did Kennedy have than Nixon?

_____

22. In 1992, Bill Clinton had 44,908,254 votes; George Bush had 39,102,343; and Ross Perot had 19,741,065 votes. Taken together, how many more votes did Bush and Perot have than Clinton?

_____

ADDING AND SUBTRACTING WHOLE NUMBERS

If you think the answer should be more than both numbers, add numbers. Otherwise, subtract.

1. Judith Stein hiked 35 miles in 2 days On Monday, she hiked 19 miles. How far did she hike on Tuesday? (hint: It is less than 35 miles, so you must sub–tract.)

    Answer: _____

2. The Music Department staged a spring concert. 427 people heard the concert on April 2. 498 people came on April 3. What was the total audience for the two nights?

    Answer: _____

Elena Romano's bowling score was 231. Angela Austin bowled 149.

3. By how many pins did Elena beat Angela?

    Answer: _____

4. What was their total combined score?

    Answer: _____

A bike usually sells for $150. This week, it is on sale for $127. A light is $17 extra.

5. Mary buys the bike on sale. How much does she save?

    Answer: _____

6. Sue buys the bike on sale. She also buys a light. How much does she spend?

    Answer: _____

Last Friday night, Zach Shooter scored 27 points, Freddie Frenzie scored 19 points, and Ty Dupp scored 12 points.

7. How many points did they score all together?

    Answer: _____

8. How many more points did Zack score than Ty?

    Answer: _____

Rocky Ramirez was a baseball player for 2 years.
The first year, he had 98 hits. The second year, he had 129 hits.

9. How many hits did he have in 2 years?

10. How many more hits does he need to have 1000.

    Answer: _____

    Answer: _____

USING ORDER OF OPERATIONS

If there are no parentheses, add or subtract from left to right. If there are parentheses, do what is inside the parentheses first.

EXAMPLES

$33 - 12 + 3$
$21 + 3$
24

$19 - (8 - 3)$
$19 - 5$
14

Add or subtract as indicated.

1. $52 - 6 - 18 = $ _____

2. $9 + 26 - 13 = $ _____

3. $23 - 10 + 22 = $ _____

4. $35 - (16 - 4) = $ _____

5. $45 + 6 + 19 = $ _____

6. $6 + 53 - 9 = $ _____

7. $59 - 10 + 8 = $ _____

8. $(13 + 76) - 9 = $ _____

9. $73 + 67 - 29 = $ _____

10. $340 - 5 + 81 = $ _____

11. $73 - 6 - 3 = $ _____

12. $55 - (83 - 44) = $ _____

13. $16 + 32 + 5 = $ _____

14. $(41 - 6) + 91 = $ _____

15. $87 - 8 + 23 = $ _____

16. $54 - (26 - 8) = $ _____

17. $48 + 29 + 3 = $ _____

18. $61 - 9 - 25 = $ _____

19. $714 + 7 + 60 = $ _____

20. $94 + 7 - 91 = $ _____

21. $71 - (41 - 9) = $ _____

8

MULTIPLICATION

Complete the multiplication table. Some numbers are filled in to help you get started.

1	2	3	4	5	6	7	8	9	10	11	12
2	4	6	8	10							
3	6	9	12								
4	8										
5							40				
6											
7					42						
8								72			
9											
10											
11											
12											

MULTIPLYING WHOLE NUMBERS

When you multiply, you will usually have to carry to the next column.

EXAMPLES
$$\begin{array}{r} 230 \\ \times\ 3 \\ \hline 690 \end{array}$$
$$\begin{array}{r} {}^{4} \\ 27 \\ \times\ 6 \\ \hline 162 \end{array}$$ ← Since 6 x 7 gives a two-digit number, you have to carry.

Multiply.

1. 10 $\times 2$	**2.** 14 $\times 2$	**3.** 23 $\times 3$	**4.** 30 $\times 3$	**5.** 21 $\times 4$	**6.** 11 $\times 8$
7. 73 $\times 2$	**8.** 44 $\times 2$	**9.** 93 $\times 3$	**10.** 71 $\times 4$	**11.** 53 $\times 2$	**12.** 71 $\times 7$
13. 931 $\times\ 3$	**14.** 743 $\times\ 2$	**15.** 911 $\times\ 5$	**16.** 720 $\times\ 4$	**17.** 613 $\times\ 3$	**18.** 514 $\times\ 2$
19. 24 $\times 3$	**20.** 76 $\times 5$	**21.** 29 $\times 8$	**22.** 34 $\times 9$	**23.** 38 $\times 8$	**24.** 46 $\times 4$
25. 65 $\times 3$	**26.** 49 $\times 5$	**27.** 57 $\times 8$	**28.** 56 $\times 9$	**29.** 57 $\times 7$	**30.** 75 $\times 4$

31. A jacket costs $49. How much do 7 similar jackets cost?

_____

32. A TV costs $865. How much will 8 TVs of the same model cost?

_____

33. A computer costs $789. How much do 9 similar computers cost?

_____

34. Pat charges $45 to clean a house. What will it cost to have the house cleaned 6 times?

_____

35. Two tables cost $262. How much will 9 of the same type table cost?

_____

36. A work crew earns $287 an hour. How much will it earn in 7 hours?

_____

MULTIPLYING WHOLE NUMBERS

$$
\begin{array}{r}
26 \\
\times 38 \\
\hline
208 \\
780 \\
\hline
988
\end{array}
$$

208 ⟵ 8 x 26

780 ⟵ 30 x 26   Write 0 in the ones place. Then multiply 3 x 26.

Multiply.

1. 25 x90	**2.** 56 x59	**3.** 48 x92	**4.** 94 x22	**5.** 45 x93	**6.** 59 x73
7. 578 x 20	**8.** 436 x 59	**9.** 739 x 64	**10.** 476 x 87	**11.** 324 x 68	**12.** 748 x 96
13. 8462 x 53	**14.** 5648 x 72	**15.** 7090 x 46	**16.** 8083 x 93	**17.** 3654 x 25	**18.** 4036 x 64
19. 60,008 x 86	**20.** 36,724 x 57	**21.** 23,786 x 56	**22.** 80,591 x 70	**23.** 36,428 x 93	**24.** 29,467 x 68

25. A circus calculates that the average weight of its 7 giraffes is 3,764 lbs. What is the total weight of the 7 giraffes?

26. A particular concrete building block weighs 37 lbs. How much do 2,458 such building blocks weigh?

27. If one seal can dive to a depth of about 900 ft, to what depth can 3 similar seals dive?

28. A particular thoroughbred horse weighs about 500 kg. About how much would 25 similar thoroughbreds weigh?

29. Whitney Johnson was paid $2,692 per month for three years. What were her total earnings for that time?

30. Fernando Alvarez paid $4,709 per acre for 36 acres of land. How much did he pay in all?

MULTIPLYING WHOLE NUMBERS

Multiply.

1. 800 x 10 = _____

2. 10 x 100 = _____

3. 100 x 300 = _____

4. 80 x 100 = _____

5. 32 x 10,000 = _____

6. 63 x 100 = _____

7. 10,000 x 9 = _____

8. 100 x 10,000 = _____

9. 100 x 90 = _____

10. 10 x 10,000 = _____

11. 10,000 x 68 = _____

12. 2000 x 10,000 = _____

13. 300 x 1000 = _____

14. 1000 x 7000 = _____

15. 200 x 10 = _____

16. 74 x 100 = _____

17. 50 x 1000 = _____

18. 4000x10= _____

19. 25 x 100 = _____

20. 30 x 100 = _____

21. 2000 x 10 = _____

22. 25 x 1000 = _____

23. 5 x 100 = _____

24. 10,000 x 10 = _____

25. 1000 x 40 = _____

26. 6000 x 100 = _____

27. 8 x 1000 = _____

28. 400 x 100 = _____

29. 4 x 1000 = _____

30. 1000 x 6000 = _____

31. 4 x 6000 = _____

32. 50 x 3 = _____

33. 30 x 7000 = _____

34. 4 x 900 = _____

35. 7000 x 9 = _____

36. 800 x 2 = _____

---

37. If an average cow weighs 600 kg, about how much will 60 average cows weigh?

_____

38. The range of weight for a Doberman Pinscher is 60 to 75 lb. What is the least and greatest total amount that 100 Doberman Pinscers should weigh?

_____

39. The Office Store sold 200 computers for $400 each. What was the total amount of money earned from the sale?

_____

40. Lucy Wu saved $20 a week for 200 weeks. How much had she placed in savings in that time?

_____

EXPONENTS

E X A M P L E S $2^4 = 2 \times 2 \times 2 \times 2$ $3^3 = 3 \times 3 \times 3$

$\underbrace{4} \times 2$ $9 \times 3 = 27$

$8 \times 2 = 16$

Raise to the power shown. Use separate work paper if necessary.

1. $2^2 = \underline{\quad 4 \quad}$
$2 \times 2 = 4$

2. $8^2 = \underline{\qquad}$

3. $9^2 = \underline{\qquad}$

4. $3^3 = \underline{\qquad}$

5. $4^3 = \underline{\qquad}$

6. $2^3 = \underline{\qquad}$

7. $6^2 = \underline{\qquad}$

8. $13^2 = \underline{\qquad}$

9. $15^2 = \underline{\qquad}$

10. $1^3 = \underline{\qquad}$

11. $6^3 = \underline{\qquad}$

12. $7^3 = \underline{\qquad}$

13. $2^4 = \underline{\qquad}$

14. $3^4 = \underline{\qquad}$

15. $1^4 = \underline{\qquad}$

16. $9^3 = \underline{\qquad}$

17. $10^3 = \underline{\qquad}$

18. $12^2 = \underline{\qquad}$

19. $2^6 = \underline{\qquad}$

20. $3^5 = \underline{\qquad}$

21. $2^9 = \underline{\qquad}$

22. $9^4 = \underline{\qquad}$

23. $23^2 = \underline{\qquad}$

24. $11^3 = \underline{\qquad}$

25. $5^3 = \underline{\qquad}$

26. $10^6 = \underline{\qquad}$

27. $100^2 = \underline{\qquad}$

28. $8^1 = \underline{\qquad}$

29. $20^3 = \underline{\qquad}$

30. $6^4 = \underline{\qquad}$

---

31. If 7 girls each own 7 dresses and each of the 7 dresses has 7 different hats to go with it, how many hats are there?

_____

32. Write an expression with a number raised to a power for the situation described in Exercise 31.

_____

33. If 20 schools each have 20 classrooms and each classroom has 20 desks, how many desks are there in all?

_____

34. Write an expression with a number raised to a power for the situation described in Exercise 33.

_____

DIVIDING WHOLE NUMBERS

REMEMBER Division is the reverse of multiplication. To do division, you must know all your multiplication facts.

EXAMPLES **56 ÷ 8 = 7 because 7 x 8 = 56**
36 ÷ 6 = 6 because 6 x 6= 36

Write each answer. Then write the multiplication fact that shows that the answer is right. The first few exercises are partly worked out for you.

1. 45 ÷ 5 = ___9___   5 x 9 = 45

2. 12 ÷ 4 = ____   _____

3. 25 ÷ 5 = ___5___   5 x 5 _____

4. 15 ÷ 5 = ____   _____

5. 40 ÷ 8 = ___5___   8 x 5 _____

6. 16 ÷ 4 = ____   _____

7. 32 ÷ 4 = ___8___   4 x _____

8. 28 ÷ 7 = ____   _____

9. 63 ÷ 7 = ____   _____

10. 48 ÷ 6 = ____   _____

11. 28 ÷ 4 = ____   _____

12. 49 ÷ 7 = ____   _____

13. 20 ÷ 5 = ____   _____

14. 54 ÷ 9 = ____   _____

15. 27 ÷ 3 = ____   _____

16. 64 ÷ 8 = ____   _____

17. 30 ÷ 6 = ____   _____

18. 8 ÷ 2 = ____   _____

19. 35 ÷ 7 = ____   _____

20. 72 ÷ 8 = ____   _____

21. 54 ÷ 9 = ____   _____

22. 81 ÷ 9 = ____   _____

23. 35 ÷ 7 = ____   _____

24. 56 ÷ 7 = ____   _____

25. 42 ÷ 6 = ____   _____

26. 63 ÷ 9 = ____   _____

27. $56 is divided evenly among 8 people. How much does each person receive?

_____

28. A dinner bill for $81 is divided among 9 people. How much does each person pay?

_____

29. If 40 people sit for dinner with 8 people at a table, how many tables are needed?

_____

30. If 20 people have $2 each, how much money do they have altogether?

_____

14

DIVIDING WHOLE NUMBERS

$\frac{21}{8)168}$ How many 8s in 16? __2__   How many 8s in 8? __1__

Divide.

1. $3)\overline{9}$ **2.** $4)\overline{8}$ **3.** $6)\overline{6}$ **4.** $2)\overline{8}$ **5.** $3)\overline{6}$

6. $1)\overline{7}$ **7.** $4)\overline{28}$ **8.** $9)\overline{27}$ **9.** $3)\overline{24}$ **10.** $8)\overline{56}$

11. $7)\overline{63}$ **12.** $5)\overline{35}$ **13.** $2)\overline{84}$ **14.** $4)\overline{40}$ **15.** $3)\overline{39}$

16. $2)\overline{68}$ **17.** $5)\overline{55}$ **18.** $3)\overline{96}$ **19.** $7)\overline{490}$ **20.** $6)\overline{366}$

21. $4)\overline{288}$ **22.** $2)\overline{128}$ **23.** $3)\overline{189}$ **24.** $8)\overline{168}$ **25.** $2)\overline{182}$

26. $3)\overline{183}$ **27.** $5)\overline{150}$ **28.** $9)\overline{5409}$ **29.** $3)\overline{2196}$ **30.** $2)\overline{1426}$

---

31. If 3 identical cars weigh a total of 6930 pounds, how much is the weight of each car?

32. If $729 is divided evenly among 9 people, how much does each person receive?

33. If a plane travels 3507 miles in 7 hours, what is its average speed per hour?

34. If a car travels an average speed of 45 miles an hour, how far will it travel in 5 hours?

35. A 126–page book contains 6 chapters. Each chapter has the same number of pages. How many pages does each chapter contain?

36. A $1505 prize is offered for the solution to a puzzle. Five people submit the correct solution. How much does each person receive if they share the prize equally?

DIVIDING WHOLE NUMBERS

$$\begin{array}{r} 17 \\ 5\overline{)85} \\ 5 \\ \hline 35 \\ 35 \\ \hline 0 \end{array}$$

How many 5s in 8? **1**

Multiply 1 x 5. Subtract. **Bring down the next number.**

How many 5s in 35? **7**

Multiply 7 x 5. Subtract.

0 ⟵ There is no remainder.   **Answer:17**

Divide.

1. $3\overline{)57}$ **2.** $5\overline{)85}$ **3.** $4\overline{)68}$ **4.** $2\overline{)94}$ **5.** $7\overline{)84}$ **6.** $8\overline{)96}$

7. $6\overline{)72}$ **8.** $3\overline{)78}$ **9.** $5\overline{)70}$ **10.** $4\overline{)92}$ **11.** $6\overline{)96}$ **12.** $3\overline{)87}$

13. $9\overline{)657}$ **14.** $5\overline{)430}$ **15.** $6\overline{)594}$ **16.** $7\overline{)476}$ **17.** $8\overline{)232}$ **18.** $4\overline{)136}$

19. $2\overline{)578}$ **20.** $3\overline{)852}$ **21.** $4\overline{)916}$ **22.** $5\overline{)865}$ **23.** $6\overline{)768}$ **24.** $4\overline{)728}$

25. The Esposito family travels 948 miles in 4 days. What is their average distance traveled per day?

26. If $1099 is divided evenly among 7 house painters, how much is each painter paid?

27. A party costing $60 is paid for by 6 people. Each person pays an equal share. How much does each person pay?

28. A person gives $50 to each of 5 charities. How much does he give away in all?

16

DIVIDING WHOLE NUMBERS

```
     253
7)1771    How many 7s in 17?  2
   14     Multiply 2 x 7.  Subtract.  Bring down the next number.
   37     How many 7s in 37?  5
   35     Multiply 5 x 7.  Subtract.  Bring down the next number.
   21     How many 7s in 21?  3
   21     Multiply 3 x 7.  Subtract.
    0     There is no remainder.      Answer: 253
```

Divide.

1. 5)2970   **2.** 8)2952   **3.** 2)1396   **4.** 7)8792   **5.** 6)1470

6. 4)1428   **7.** 9)1647   **8.** 8)7808   **9.** 7)5173   **10.** 6)3408

11. 9)4473   **12.** 5)6185   **13.** 3)4695   **14.** 4)6924   **15.** 3)7104

---

16. If $54,971 is divided equally among 7 people, how much does each receive?

_____

17. If $54,522 is divided equally among 9 people, how much does each receive?

_____

18. If 26,808 acres of land is divided into 6 plots that are the same size, how many acres will each lot contain?

_____

19. If a trip of 3080 miles takes 8 days, what is the average distance covered each day?

_____

DIVIDING WHOLE NUMBERS

$$\begin{array}{r} 3 \\ 21\overline{)67} \\ \underline{63} \\ 4 \end{array}$$

21 is close to 20.
How many 20s in 67?
(How many 2s in 6?) __3__
Answer 3 R4

$$\begin{array}{r} 26 \\ 25\overline{)673} \\ \underline{50} \\ 173 \\ \underline{150} \\ 23 \end{array}$$

Answer: 26 R23

Divide.

1. $31\overline{)75}$ **2.** $57\overline{)61}$ **3.** $21\overline{)88}$ **4.** $48\overline{)96}$ **5.** $16\overline{)71}$ **6.** $12\overline{)55}$

7. $23\overline{)87}$ **8.** $75\overline{)93}$ **9.** $11\overline{)70}$ **10.** $29\overline{)80}$ **11.** $25\overline{)78}$ **12.** $14\overline{)98}$

13. $21\overline{)735}$ **14.** $51\overline{)238}$ **15.** $35\overline{)630}$ **16.** $23\overline{)492}$ **17.** $37\overline{)803}$

18. $17\overline{)620}$ **19.** $19\overline{)504}$ **20.** $28\overline{)816}$ **21.** $38\overline{)743}$ **22.** $47\overline{)624}$

23. If \$243 is collected from the sale of 27 tickets, what is the cost of each ticket?

24. If 280 chairs are arranged into 35 rows, how many chairs are in each row?

25. If 240 containers of juice are delivered to the school each day, how many are delivered in three days?

26. If a 296 mile biking trip is covered in 10 days, how many miles are biked each day, to the nearest mile?

WRITING WHOLE NUMBERS

millions   ten thousands   hundreds   ones

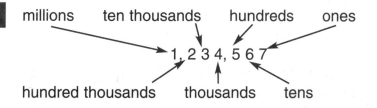

hundred thousands   thousands   tens

EXAMPLES   **8,132,095 = eight million one hundred thirty–two thousand**

Write each number in words.

1. 4163 _____

2. 2,500,120 _____

3. 60,905 _____

4. 83,407,600 _____

5. 49,038 _____

6. 303,000 _____

7. 2906 _____

8. 4,902,000 _____

9. 63,018 _____

10. 38,000,100 _____

11. 520,006 _____

12. 100,010,000 _____

Write each number in digits.

13. Nine hundred forty _____

14. Three thousand seventy–two _____

15. Six hundred eight million _____

16. Nineteen thousand six hundred _____

17. Sixty–five hundred _____

18. Thirty thousand eight _____

19. Four million one hundred twenty–seven _____

20. Seventy thousand three hundred fifty–nine _____

ROUNDING WHOLE NUMBERS

If the digit to the right of the rounding place is 5 or more, round to the next higher digit. Otherwise, round down.

EXAMPLES **Round 5263 to the nearest hundred.**   hundreds

5263

The digit to the right of the hundreds place is 6. Since 6 is more than 5, round up.   Answer: 5300

Round to the nearest ten.

1. 58 _____

2. 2094 _____

3. 16 _____

4. 125 _____

Round to the nearest hundred.

5. 230 _____

6. 9508 _____

7. 942 _____

8. 1373 _____

Round to the nearest thousand.

9. 9158 _____

10. 13,814 _____

11. 4039 _____

12. 980 _____

Round to the nearest ten thousand.

13. 8,541,396 _____

14. 48,302,767 _____

Round to the nearest hundred thousand.

15. 8,541,396 _____

16. 48,302,767 _____

Listed below in feet are the highest points in several states. Round each height to the nearest thousand feet.

17. Mt. McKinley   20,320 _____

18. Boundary Peak   13,140 _____

19. Mt. Whitney   14,494 _____

20. Mt. Washington   6,288 _____

21. Mouna Kea   13,796 _____

22. Mt. Hood   11,235 _____

23. Mt. Rainier   14,410 _____

24. Mt. Marcy   5,344 _____

The data below are from the official U.S. population census. Round each number to the nearest ten thousand.

25. 1790   3,929,214 _____

26. 1910   91,972,266 _____

27. 1830   12,866,020 _____

28. 1940   131,669,275 _____

29. 1870   38,558,371 _____

30. 1970   203,235,298 _____

ORDER OF OPERATIONS

In doing the four basic operations, you follow these rules.
(1) Do what is in parentheses, if there are parentheses.
(2) Do multiplication and division from left to right.
(3) Do addition and subtraction from left to right.

EXAMPLES

$5(3 + 7) - 13$
$5(10) - 13$
$50 - 13$
37

$7(9-5) + 18 \div (7-4)$
$7(4) + 18 \div 3$
$28 + 6$
34

Simplify by performing the operations in the proper order.

1. $7(2+7) - 15$

2. $8 + 2(9-2)$

3. $6(7 - 1) + (3 + 11)$

4. $45 \div (9 + 6) - 1$

5. $11 + (6 \div 2) - 7$

6. $4(12 - 5) - 13$

7. $(32 + 4) \div 9 + 10$

8. $5 \times 5 - 3 \times 2$

9. $11(9-5) - (8 \div 2)$

10. $4(28-21) + 3$

11. $17 - 3(19 - 14)$

12. $4 \times 13 + 51 \div 3$

13. $35 \div (12 - 7) \times 2$

14. $3 + 15 \times 2 - 11$

15. $9(2 + 9) - 5 \times 8$

16. $6(44 - 21) - 55$

17. $8 + 3(12 + 11)$

18. $6(5 + 5) + (45 - 13)$

EVALUATING MATHEMATICAL EXPRESSIONS

REMEMBER: To evaluate a mathematical expression you substitute a particular number for a letter and then do the operations.

EXAMPLES **Evaluate when $a = 2$.**
$2(a + 7) + 3$
$2(2 + 7) + 3$
$2 \times 9 + 3$
$18 + 3$
21

Evaluate when $a = 3$; $a = 5$.
$8b - (a + 4)$
$8 \times 5 - (3 + 4)$
$40 - 7$
33

Evaluate each expression for $a = 2$, $b = 5$, $c = 8$.

1. $6a + 5$

2. $4b - 1$

3. $a + b$

4. $(7 + 4) - c$

5. $b(3 + 6)$

6. $(9 - a) + (10 \div 2)$

7. $3b - c$

8. $(c + 12) \div 5$

9. $a + b + c$

10. $7a + 2b$

11. $3(a + b)$

12. $9(7 - a)$

13. $(6 + c) \times 2$

14. $6(c - a)$

15. $9b - 3 \times 6$

16. $5b - 10 + 2c$

17. $2(a + b) - 7$

18. $7c - b(5 - 2)$

22

DIVISION: PRACTICAL APPLICATIONS

Read the problems carefully. It is not always correct to round to the nearest whole number.

EXAMPLES **72 students signed up for a field trip. Each bus holds 30 students. How many buses will be needed?**

$$2\frac{12}{30} \text{ or } 2\frac{2}{5}$$
Divide. $30\overline{)72}$

If you round $2\frac{2}{5}$ you get 2. However, you need 3 buses, not 2.

1. If tickets for a basketball game cost $3 each, how many people could attend the game for $200?

   Answer: _____

2. Lee and Chris hiked for 12 days, covering 140 miles. How many miles per day did they average?

   Answer: _____

3. Martin is buying hot dogs for a picnic for 19 people. Each package contains enough hot dogs for 3 people. How many packages should he buy?

   Answer: _____

4. In the 12 years from 1920 to 1931, Babe Ruth hit 562 home runs. How many homers did he average per year?

   Answer: _____

5. Sarah's bowling team has 3 members. If they buy a jumbo pizza having 16 slices, how many whole slices should each person get?

   Answer: _____

6. Fred is planning a trip from Ohio to Idaho, a total of 2195 miles. If he drives 450 miles per day, how many days will the trip take?

   Answer: _____

7. If Fred in Exercise 6 makes the trip from Idaho to Ohio in 4 days, how many miles must he drive per day?

   Answer: _____

DECIDING WHEN TO ESTIMATE

REMEMBER: Look to see whether an estimate is enough to solve a problem.
An estimate is usually easier to find than an exact answer.

EXAMPLES **Asa is driving 135 miles to see a basketball game. The game starts in 3 hours. If she drives 55 miles per hour, will she arrive in time?**

Use 50 miles per hour to estimate.
3 hours x 50 miles per hour = 150 miles
Asa can drive about 150 miles in 3 hours. She will arrive on time.

---

1. Mr. Lee's car gets 22 miles per gallon of gas. The gas tank holds 14 gallons. Is one tank of gas enough to drive 250 miles?

Answer: _____

2. Max has a bowling average of 189. He needs to bowl 612 for 3 games for his team to win. If he bowls his average, will they win?

Answer: _____

---

3. On his daily commute, Mr. Lee pays 6 tolls that vary from $0.75 to $1.25. He works 20 days each month. About how much does he pay in tolls each month?

Answer: _____

4. In his first two seasons playing baseball, Omar had 209 hits each year. If he averages 209 hits in future years, how many years will it take for him to get 3,000 hits?

Answer: _____

---

5. Tennis balls cost $3.95 a can. How many cans can Sharon buy for $20?

Answer: _____

6. Kisha has a job offer that pays $7.89 per hour for 40 hours per week. She needs to earn $400 per week. Does the job pay enough?

Answer: _____

---

7. Ms. Carr reads about 35 to 40 pages per hour. If she reads 1 hour each day, can she finish a 350–page book in one week?

Answer: _____

8. Mr. Williams often flies from home to Omaha and back. The total distance is 2,512 miles. How many trips must he make to earn a free ticket for flying 25,000 miles?

Answer: _____

9. Siva can only afford $8000 a year for rent. Which of these apartments could she afford to rent?

Answer: _____

Apartments for Rent	
Studio	$450 per month
1-Bedroom	$600 per month
2-bedroom	$800 per month

10. One night, Ralph ate an entire 12-ounce bag of chips. Was that more than 1000 calories?

Answer: _____

Food	Calories
Potato chips	150 (1 ounce)
Grapefruit	78 (1 grapefruit)
Deluxe burger	500
Celery	10 (1 stalk)

11. Which has more calories, 1 deluxe burger or 5 grapefruits?

Answer: _____

12. One night, Ralph ate an entire bunch of 12 celery stalks. Was that more than 1000 calories?

Answer: _____

CHAPTER 1

POSTTEST

Add.

1. 62
   +35

2. 93
   + 0

3.   6
    +8

4. 74
   +19

5. 235
   +687

6. 29
    13
   +66

7. 83 + 259 = ____

Subtract.

8.   96
    −30

9.   64
    −18

10. 70
    −38

11. 501
    −213

12. 736
    −539

13. 87 − (45 − 11) = ____

Multiply.

14. 23
    x 3

15. 47
    x 2

16. 68
    x 5

17. 719
    x 52

18. 408
    x 19

19. 41 x 26 = ____

Divide.

20. 3)93

21. 4)420

24. 12 ÷ 3 − 1

25. 8(5+2) − 3

22. 14)476

23. 35)762

26. Evaluate 3(x + y) for x = 3, y = 4

Raise to the power shown.

27. 6^3 = _____

28. 7^2 = _____

Round to the nearest thousand.

29. 1342 _____

30. 2583 _____

31. 19,720 _____

Write the numbers.

32. Four hundred sixty-two thousand five _____

33. Two hundred three _____

PRETEST

Add.

1. $1.50
     + 4.25

2. $34.57
      42.46
     + 56.79

3. 62.84
      1.9731
     +15.6

4. $9 + 6.38 + .175 =$ _____

Subtract.

5. $35.78
     − 2.34

6. $62.75
     − 9.89

7. 4.23
     −1.085

8. $6.21 − 4.3 =$ _____

Multiply.

9. $13.49
     x   6

10. $54.63
      x   86

11. 92.1
      x 4.6

12. $2.75 \times 3.8 =$ _____

Divide.

13. $8\overline{)19.2}$

14. $15\overline{)\$63.75}$

15. $2.1\overline{)4.41}$

16. $5.50 \div 0.8 =$ _____

Round $673.524 to the nearest:

17. Dollar _____

18. Hundred dollars _____

19. Cent _____

Work out the following exercises.

20. Add $35.16, $62.09, and $46. _____

21. Subtract $5.09 from $20. _____

22. Multiply fourteen dollars ten cents by seventy-four. _____

Solve each equation.

23. $a + 15 = 60$

24. $3.5 = b − 1.5$

25. A hat costs $10.50 and a shirt costs $9.95. Find the cost of both.

Answer: _____

26. If you bought a book for $6.75, what would the change from $10 be?

Answer: _____

27. A chair costs $83.80. Find the cost of 4 chairs.

Answer: _____

28. A burger costs $2.29 and a soft drink costs $0.69. Find the cost of 2 burgers and 3 drinks.

Answer: _____

29. If you earned $60 in 8 days, how much money was that for each day?

Answer: _____

30. If you paid $12.15 for 9 gallons of gas, what was the cost per gallon?

Answer: _____

31. If you save $4 per week, how long will it take you to save $240?

Answer: _____

32. If you earn $85 per week, how much will you earn in 13 weeks?

Answer: _____

33. Circle the largest number.   0.83   0.9   0.806

28

PLACE VALUE OF DECIMALS

tens   tenths   thousandths   hundred thousandths

2 3 . 5 6 8 2 7

ones   hundredths   ten thousandths

EXAMPLE   **Write the number "fifty-three hundredths."**
The number ends in hundredths place. So there are two decimal places.
Answer: 0.53

Write each number in digits.

1. Six tenths _____

2. Forty-nine hundredths _____

3. Eight hundredths _____

4. Ninety hundredths _____

5. One and five tenths _____

6. Six hundred thousandths _____

7. Seven hundred eighty-three thousandths _____

8. Four and seventy-seven thousandths _____

Give the place value of the 6 in each number.

EXAMPLE   **0.5613   Answer: hundredths**

9. 0.638 _____

10. 328.60 _____

11. 0.9601 _____

12. 16.91 _____

13. 2.306 _____

14. 613.5 _____

15. 6.9 _____

16. 8.962 _____

17. 0.0006 _____

18. 0.306 _____

19. 0.31906 _____

20. 0.0906 _____

21. 58.62 _____

22. 1.0976 _____

23. 918.06 _____

24. 0.034461 _____

25. 0.35060 _____

26. 0.9162 _____

COMPARING DECIMALS

You can add zeros to the right side of a decimal without changing its value.

EXAMPLE   **Of 0.34 and 0.304, circle the larger number.**
0.340   **The two decimals have different numbers of places.**   **(0.34)**
0.304   **Add a zero to the 0.34, and then compare them.**   **0.304**

Circle the larger number in each pair. If the two numbers are equal, circle both.

1. 0.79 0.82   **2.** 0.681 0.659   **3.** 0.318 0.254

4. 0.58 0.6   **5.** 0.21 0.2   **6.** 0.60 0.6

7. 0.803 0.308   **8.** 0.030 0.003   **9.** 0.206 0.26

10. 0.930 0.93   **11.** 0.8 0.888   **12.** 0.5 0.381

13. 0.407 0.704   **14.** 0.061 0.61   **15.** 0.301 0.31

16. 0.65 0.617.   **17.** 0.93 0.930   **18.** 0.206 0.026

19. 0.189 0.19   **20.** 0.6 0.600   **21.** 0.7 0.077

22. 0.618 0.6180   **23.** 0.406 0.604   **24.** 0.91 0.290

25. 0.9 0.869   **26.** 0.05 0.494   **27.** 0.16 0.9

28. 2.3 0.23   **29.** 0.9 8.8   **30.** 2.1 2.08

EXAMPLE   **Write 0.27, 0.207, 0.7, and 0.72 in order from largest to smallest.**

0.270   **First line up the numbers vertically with the decimal points**
0.207   **one under another. Then add zeros on the right until all the**
0.700   **numbers have the same number of digits.**
0.720   **Answer: 0.72,   0.7,   0.27,   0.207**

List the numbers in order, from largest to smallest.

31. 0.24,   0.2,   0.42,   0.204   _____

32. 0.608,   0.86,   0.086,   0.9   _____

33. 0.307,   0.70,   0.703,   0.37   _____

30

COMPARING DECIMALS

Zeros after a whole number make it larger: 100 is larger than 10. Zeros after a decimal point but before the first digit make the number smaller: 0.01 is smaller than 0.10. Zeros after the digits in a decimal do not change the value: 0.10 and 0.100 are equal.

EXAMPLE   Circle the largest number.   3.00   0.300   0.03   (30.0)
           Circle the smallest number.   (0.055)   0.55   5.5   0.50

Circle the largest number.

1. 0.90   0.09   9.0

2. 0.09   0.08   0.93

3. 0.1   0.12   0.21

4. 1.0   10   0.01

5. 0.64   0.603   0.6

6. 8.01   8.11   8.0

7. 60   60.6   6.66

8. 3   0.38   0.308

9. 4.8   4.5   3.7

10. 0.77   0.077   0.007

11. 1.7   1.11   1.17

12. 10.1   11.0   11.01

13. 0.04   0.05   0.06

14. 90   9.6   9.35

15. 2.2   2.12   2.02

16. 2.3   0.32   3.2

17. 0.09   8   3.95

18. 1.0   1.9   1.2

19. 0.05   0.5   5.6

20. 6.4   3.08   7

21. 9.5   9.4   9.85

Circle the smallest number.

22. 0.90   0.09   9.0

23. 9.06   6.9   9.6

24. 0.5   0.4   0.05

25. 1.0   10   0.01

26. 0.07   0.4   0.43

27. 1.0   2.1   1.21

28. 0.2   0.22   0.222

29. 3.19   3.9   3

30. 5.4   5.39   5.41

31. 0.08   0.88   0.80

32. 380   83   83.5

33. 18   8.1   17.99

34. 3.3   33   0.33

35. 0.407   0.47   0.7

36. 0.01   0.001   0.1

37. 54   540   5.4

38. 0.03   0.3   0.003

39. 25   25.1   26

40. 6.6   6.06   66

41. 5.9   3   0.95

42. 10   9   8.8

43. 4.9   0.49   .409

44. 0.061   0.601   0.610

45. 6.05   6.5   6.6

46. 1.7   1.07   0.71

47. 0.65   0.9   3.1

48. 99   97   9.8

ADDING MONEY

Line up the decimal points, so that you add dollars to dollars and cents to cents.

EXAMPLE	$2.25	$6.83
	+ 3.41	+ 7.97
	$5.66	$14.80

Add

1. $2.50
   + 3.42

2. $1.61
   + 7.14

3. $2.25
   + 4.53

4. $3.87
   + 4.12

5. $9.61
   + .25

6. $3.07
   + 5.91

7. $12.34
   + 7.25

8. $25.67
   + 10.32

9. $27.14
   + 1.05

10. $91.37
    + 6.50

11. $40.09
    + 51.80

12. $ 2.46
    + 32.13

13. $4.72
    + 3.19

14. $5.63
    + 2.17

15. $9.28
    + .36

16. $10.56
    + 9.36

17. $21.39
    + 34.55

18. $46.29
    + 13.14

19. $7.72
    + 6.26

20. $8.53
    + 1.19

21. $4.45
    + 4.73

22. $23.65
    + 9.04

23. $2.09
    + 30.19

24. $18.67
    + 4.12

25. Cooper purchases groceries for $37.89 and cleaning materials for $9.87. How much does he spend?

26. Kelly buys a dress for $34.56 and a sweater for $18.07. What is the total amount of her bill?

27. Shawn spends $20 for gas and $13.45 for dinner. What is the total for the two purchases?

28. Lily buys a video for $19.95 and film for $8.75. How much does she spend?

29. Michael eats a meal costing $18.86. What is his change from $20?

30. Kendra spends $187.68 for skis and $88.35 for a jacket. What is the total amount that she spends?

ADDING MONEY

Watch for numbers adding up to 10: 8 + 2, 7 + 3, 6 + 4

EXAMPLE
$$\begin{array}{r} \$\,9.57 \\ 23.24 \\ +\,81.69 \\ \hline \$114.50 \end{array}$$

Add the following amounts of money. Read each amount to yourself.

1. $11.23
    3.34
    + 10.21

2. $41.72
    23.13
    + 34.02

3. $20.00
    3.00
    + 0.08

4. $20.01
    5.90
    + 12.08

5. $100.50
    75.05
    +   4.33

6. $34.50
    44.76
    + 21.34

7. $67.38
    41.22
    + 9.03

8. $49.09
    1.01
    + 56.62

9. $51.62
    9.08
    + 90.55

10. $203.45
     73.65
     + 101.99

11. $11.23
     23.25
     31.24
     + 45.42

12. $23.45
     45.36
     32.16
     + 53.24

13. $56.54
     24.63
     57.47
     + 10.09

14. $67.87
     25.94
     81.66
     + 25.55

15. $404.99
     75.01
     405.44
     + 40.86

16. Mr. Ramirez buys a TV for $356 and a VCR for $169.95. How much does he spend for the two items?

17. Mrs. Wilson buys a computer for $1209 and software for $119.76. What is the cost of the computer and software together?

18. Jamie Metzer earns $125.50 on Monday, $165.49 on Tuesday, and $173.20 on Wednesday. How much does she earn for the three days?

19. John Lenov paid his workers: $456.78, $535.90, and $602.82. How much did he pay altogether?

ADDING MONEY

Read carefully. Write the numbers, lining up decimal points.

EXAMPLE **Add: Four dollars and ten cents** $4.10
             Three dollars and twenty-five cents   3.25
             Seven dollars and eleven cents   + 7.11
                                          $14.46

Add the amounts of money.

1. Five dollars and thirty-five cents
    Eight dollars and fourteen cents
    Nine dollars and forty-six cents

    Answer: _____

2. Ninety-six cents
    Four dollars and seven cents
    Twenty dollars

    Answer: _____

3. Seventy-five dollars and seven cents
    Eighty dollars and ninety-nine cents
    Sixty-five cents

    Answer: _____

4. Six cents
    Twenty-nine dollars
    Seventeen dollars and two cents

    Answer: _____

5. Sixty-two dollars and fifteen cents
    Eight-one dollars and thirty cents
    Seven dollars and eighteen cents

    Answer: _____

6. Eleven dollars and eight cents
    Fifty dollars and nine cents
    Two hundred dollars

    Answer: _____

7. Twenty dollars and forty cents
    Twelve dollars and seven cents
    Six dollars and eighty-two cents

    Answer: _____

8. Ninety-six cents
    Eighty-seven dollars
    Twenty-four dollars and six cents

    Answer: _____

ADDING MONEY: APPLICATIONS

REMEMBER: Read carefully, and label each amount.

EXAMPLE **A hat costs $6.50 and shoes cost $37.25 What is the total cost for the hat and shoes together?**

Hat	$ 6.50
Shoes	37.25
Total	$43.75

1. A shirt costs $5 and a belt costs $4.89. What is the total cost of the two together?

Shirt $5.00
Belt

Answer: _____

---

2. A bat costs $14.75 and a glove costs $27.95. How much do they cost together?

Answer: _____

3. A blouse costs $7.95, a belt costs $5.60, and shoes cost $15.40. How much do all three cost?

Answer: _____

---

4. A burger costs $1.25, a soft drink costs $0.35, and fries cost $0.55. How much is it for just the burger and soft drink?

Answer: _____

5. A shirt costs $27.50 and a tie costs $15. How much would it be for a shirt and two ties?

Answer: _____

---

6. A calculator costs $7.98 and batteries cost $0.39 each. How much is it for a calculator and two batteries.

Answer: _____

7. A radio costs $24.50 and earphones cost $8.98. How much is it for both?

Answer: _____

ADDING MONEY: APPLICATIONS

Read carefully, and answer exactly what is asked.

A clothing store listed the following prices:

Shirts	$ 6.95	Socks	$1.99 a pair
Blouses	7.50	Hat (man's)	4.50
Jeans	12.75	Hat (woman's)	6.95
Sweaters	19.95	Belts	5.99

EXAMPLE **Find the cost of a shirt and jeans.**

Shirt	$ 6.95
Jeans	+12.75
Total	$19.70

Find the cost of each of the following purchases.

1. A shirt and a sweater _____

2. Two shirts and a man's hat _____

3. A blouse and a woman's hat _____

4. A blouse, jeans, and a belt _____

5. A belt and two pairs of socks _____

6. A sweater, a man's hat, and two belts _____

7. A sweater and a belt _____

8. A pair of socks and two sweaters _____

9. Jeans and a sweater _____

10. Three belts _____

SUBTRACTING MONEY

$29.14	$46.19 − $23.67	$46.19
− 16.29	Line up the	− 23.67
$12.85	decimal points.	$22.52

Subtract

1. $45.86
− 2.15

2. $64.99
− 3.76

3. $97.46
− 17.23

4. $76.87
− 70.63

5. $31.77
− 1.56

6. $45.67
− 2.72

7. $98.56
− 24.85

8. $59.87
− 24.29

9. $27.38
− 3.82

10. $92.56
− 29.14

11. $63.34
− 9.34

12. $91.54
− 9.25

13. $73.39
− 6.24

14. $83.07
− 1.25

15. $25.94
− 6.28

16. $10.00
− 9.75

17. $30.00
− 29.50

18. $40.00
− 35.00

19. $90.00
− 1.90

20. $70.00
− 65.35

21. $74.59
− 23.67

22. $45.89
− 9.99

23. $31.22
− 6.74

24. $83.44
− 19.80

25. $90.33
− 4.69

26. $55.46
− 50.98

27. $65.34
− 36.71

28. $10.13
− 9.18

29. $15.60
− 9.68

30. $50.05
− 45.55

31. $29.25 − $2.73 = _____

32. $74.93 − $9.46 = _____

33. $78.00 − $8.01 = _____

34. $11.11 − $7.77 = _____

35. $42.66 − $38.97 = _____

36. $72.92 − $1.95 = _____

37. Joanne earns $98.30 per week working at Orchard Nursery. From each paycheck, $23.57 is taken out for taxes and other deductions. How much is her weekly paycheck?

Answer: _____

SUBTRACTING DECIMALS

REMEMBER: Line up decimal points. Write zeros so that all numbers have the same number of places.

EXAMPLE

$$\begin{array}{r} 7.6 \\ -4.219 \\ \hline \end{array}$$

Write zeros to show
hundredths and thousandths.

$$\begin{array}{r} 7.600 \\ -4.219 \\ \hline 3.381 \end{array}$$

Subtract

1. $\begin{array}{r} 9.5 \\ -3.2 \\ \hline \end{array}$

2. $\begin{array}{r} 6.213 \\ -1.8219 \\ \hline \end{array}$

3. $\begin{array}{r} 38.9 \\ -\ 9.2 \\ \hline \end{array}$

4. $\begin{array}{r} 14.20 \\ -\ 6.81 \\ \hline \end{array}$

5. $\begin{array}{r} 3.76 \\ -\ 0.29 \\ \hline \end{array}$

6. $\begin{array}{r} 0.319 \\ -0.261 \\ \hline \end{array}$

7. $\begin{array}{r} 4.903 \\ -\ 0.318 \\ \hline \end{array}$

8. $\begin{array}{r} 4.7 \\ -3.5 \\ \hline \end{array}$

9. $\begin{array}{r} 98.0 \\ -\ 2.7 \\ \hline \end{array}$

10. $\begin{array}{r} 113.92 \\ -\ 4.61 \\ \hline \end{array}$

11. $\begin{array}{r} 6.92 \\ -1.3 \\ \hline \end{array}$

12. $\begin{array}{r} 7.521 \\ -2.37 \\ \hline \end{array}$

13. $\begin{array}{r} 3.76 \\ -1.8 \\ \hline \end{array}$

14. $\begin{array}{r} .79 \\ -.5 \\ \hline \end{array}$

15. $\begin{array}{r} 9.713 \\ -5.92 \\ \hline \end{array}$

16. $\begin{array}{r} 6.5 \\ -1.88 \\ \hline \end{array}$

17. $\begin{array}{r} 38.16 \\ -\ 6.255 \\ \hline \end{array}$

18. $\begin{array}{r} 14. \\ -\ 8.6 \\ \hline \end{array}$

19. $\begin{array}{r} 4.1 \\ -2.693 \\ \hline \end{array}$

20. $\begin{array}{r} 0.4 \\ -0.261 \\ \hline \end{array}$

21. $\begin{array}{r} 0.01 \\ -0.008 \\ \hline \end{array}$

22. $\begin{array}{r} 6. \\ -\ 0.281 \\ \hline \end{array}$

23. $\begin{array}{r} 9.83 \\ -1.6 \\ \hline \end{array}$

24. $\begin{array}{r} 4.47 \\ -\ 0.8 \\ \hline \end{array}$

25. $\begin{array}{r} 20.0 \\ -\ 0.36 \\ \hline \end{array}$

26. $4.52 - 2.01 = $ _____

27. $9.3 - 4.95 = $ _____

28. $0.26 - 0.148 = $ _____

29. $2.69 - 1.5 = $ _____

30. $0.965 - 0.38 = $ _____

31. $0.093 - 0.06 = $ _____

32. What remains if you take 0.6 from 2? _____

33. Take 3.5 from 7.5. What is the answer? _____

SUBTRACTING MONEY

Read each problem carefully. Write the numbers, lining up decimal points.

EXAMPLES **If a ball costs $2.95, find the change you would get back from a $5 bill.**

Five dollars	$5.00
Cost of ball	− 2.95
Change	$2.05

1. If a bat costs $3.79, find the change you would get back from $5.

Answer: _____

2. If you left home with $5.50 and spent $2.75 for the movies, how much money would you have left?

Answer: _____

3. If you had $2.60 and bought a hamburger for $1.09, how much money would you have left?

Answer: _____

4. If you gave the salesclerk a $10 bill for a book costing $7.89, how much change would you get back?

Answer: _____

5. If you had $3.55 and bought candy for $0.79, how much money would you have left?

Answer: _____

6. If you had saved $7.65, how much more would you need to buy a concert ticket for $12?

Answer: _____

7. If your friend has $15.36 and you have $9.78, how much more does your friend have than you?

Answer: _____

8. If you had $5.32 yesterday, and now you have $3.91, how much did you spend?

Answer: _____

9. How much more than $5.89 is $14.10?

Answer: _____

ADDING AND SUBTRACTING MONEY

Read each problem carefully. Think about what must be added or subtracted. Write each item and amount, lining up decimal points.

A sports store listed the following prices:

Bat	$14.95	Tennis racket	$49.75
Baseball	3.87	Running shoes	57.65
Football	9.66	Sweatshirt	10.45

EXAMPLES **If you buy a bat, how much change will you get from a $20 bill?**

$$\begin{array}{r} \$20.00 \\ -\ 14.95 \\ \hline \$\ 5.05 \end{array}$$

For each purchase, find the amount of change you would get from $20.

1. A baseball _____

2. A baseball and bat _____

3. Two baseballs _____

4. A football _____

For each purchase, find the amount left out of $50.

5. A tennis racket _____

6. Two footballs _____

7. Sweatshirt _____

8. Two bats _____

For each purchase, find the amount left out of $100.

9. Two sweatshirts and a pair of running shoes _____

10. A tennis racket and a sweatshirt _____

11. Two bats and 10 baseballs _____

12. Two tennis rackets _____

13. Six sweatshirts and two footballs _____

14. Running shoes and a tennis racket _____

MULTIPLYING MONEY

When multiplying an amount of money by a whole number, count off two decimal places from the right to separate the dollars from the cents.

EXAMPLES

$23.47
x   3
$70.41

$ 57.09
x   8
$456.72

Multiply.

1. $45.76
x   2

2. $86.75
x   3

3. $62.91
x   5

4. $73.68
x   6

5. $49.67
x   8

6. $55.30
x   6

7. $38.97
x   5

8. $76.84
x   7

9. $45.69
x   8

10. $91.26
x   7

11. $89.74
x   9

12. $74.95
x   7

13. $96.75
x   4

14. $27.99
x   9

15. $78.66
x   6

16. $45.09
x   4

17. $70.37
x   5

18. $35.84
x   9

19. $39.08
x   3

20. $59.58
x   8

21. $39.57
x   7

22. $59.02
x   6

23. $47.80
x   9

24. $29.38
x   5

25. $39.05
x   5

26. $77.04
x   8

27. $82.47
x   7

28. $90.58
x   6

29. $30.08
x   9

30. $16.58
x   8

31. $68.97
x   9

32. $47.34
x   4

33. $98.67
x   6

34. $37.08
x   7

35. $47.68
x   9

MULTIPLYING MONEY

$38.59	**Multiply as you would with whole numbers.**
x 47	
270 13	
1543 6	
$1813.73	**Mark off two decimal places in the answer.**

Multiply.

1. $24.68	**2.** $35.79	**3.** $91.82	**4.** $47.83	**5.** $40.09
x 35	x 46	x 29	x 77	x 81

6. $15.75	**7.** $80.67	**8.** $76.00	**9.** $24.70	**10.** $99.84
x 14	x 39	x 25	x 65	x 26

11. Multiply forty-five dollars and fifty cents by thirty-seven.

Answer: _____

12. Multiply fifty-two dollars and sixty-five cents by twenty-nine.

Answer: _____

13. Multiply twenty-nine dollars and sixteen cents by fifty.

Answer: _____

14. Multiply two hundred six dollars and seventy-eight cents by fifty-nine.

Answer: _____

MULTIPLYING DECIMALS

Multiply as with whole numbers. The number of decimal places in the product is the sum of the decimal places in the factors.

EXAMPLE
```
   1.32  ◄—— two decimal places
 x 2.9   ◄—— one decimal place
  1188
   264
  3.828 ◄—— three decimal places
```

Multiply.

1. 61.2 x 8.3	**2.** 73.4 x 0.6	**3.** 70.4 x 9	**4.** 8.5 x 3.6	**5.** 71.1 x 4.3
6. 9.74 x 23	**7.** 8621 x 0,004	**8.** 39.1 x 0.62	**9.** 14.7 x 36	**10.** 0.58 x0.21
11. 0.2139 x 5.03	**12.** 0.705 x 6.38	**13.** 0.6502 x 0.017	**14.** 3.052 x 3.17	**15.** 123 x 20.9
16. 1.34 x 0.275	**17.** 0.1501 x 23.14	**18.** 4.913 x 0.266	**19.** 18.9 x 0.702	**20.** 16,248 x 0.0005

MULTIPLYING MONEY

To find the total amount of a number of the same kind of coin, multiply the amount by the number of coins.

EXAMPLES **Find the total amount for the following.**

Five quarters	Seven dimes	Six nickels	Four half dollars
$0.25	$0.10	$0.05	$0.50
x 5	x 7	x 6	x 4
$1.25	$0.70	$0.30	$2.00

Draw lines to match the amounts.

1. Three quarters	$0.14	10. Thirteen dimes	$0.70
2. Six dimes	$1.10	11. Eight nickels	$1.30
3. Four nickels	$0.85	12. Twelve quarters	$2.50
4. Fourteen pennies	$1.75	13. Seventy pennies	$0.40
5. Five half dollars	$0.20	14. Nineteen nickels	$7.50
6. Seven quarters	$0.60	15. Twenty-five dimes	$3.00
7. Eleven dimes	$0.75	16. Thirty quarters	$0.95
8. Seventeen nickels	$2.00	17. Seventy-six dimes	$2.20
9. Four half dollars	$2.50	18. Forty-four nickels	$7.60

19. Six half dollars and seven pennies	$2.80
20. Nine quarters and eleven nickels	$4.68
21. Seven quarters and eight dimes	$8.70
22. Eighteen quarters and eighteen pennies	$3.07
23. Seven half dollars and fifty-two dimes	$2.55
24. Thirty-five dimes and thirty-five nickels	$300.00
25. One thousand quarters and one thousand nickels	$5.25

44

MULTIPLYING MONEY

Read each problem carefully. Write the item and amount. Then multiply to get the total.

EXAMPLES **A shirt costs $5.79. Find the cost of 6 shirts.**

1 shirt	$5.79
	x 6
6 shirts	$34.74

1. A belt costs $4.95. Find the cost of 5 belts.

   Answer: _____

2. A pair of shoes costs $25.60. Find the cost of 3 pairs.

   Answer: _____

3. A radio costs $41.55. Find the Cost of 4 radios.

   Answer: _____

4. A blouse costs $9.67. Find the cost of 7 blouses.

   Answer: _____

5. A television set costs $179.50. Find the cost of 10 television sets.

   Answer: _____

6. A hamburger costs $1.09. Find the cost of 9 hamburgers.

   Answer: _____

7. A chair costs $56.77. How much do 6 chairs cost?

   Answer: _____

8. A watch costs $36.95. Find the cost of 8 watches.

   Answer: _____

9. A sweater costs $18.75. Find the cost of 7 sweaters.

   Answer: _____

MULTIPLYING MONEY

Hours worked x hourly rate = gross income, or pieces worked on x rate per piece = gross income.

Complete each table. Do your work on a separate piece of paper.

Name	Hours Worked						Hourly Rate	Gross Income
	M	T	W	Th	F	Total		
Example	6	4	10	7	8	35	$10.10	$176.25
1. Williams, Grace	7	8	8	8	9	_____	$9.00	_____
2. Diaz, Carlos	8	9	5	7	8	_____	$8.56	_____
3. Roberts, David	7	6	4	6	5	_____	$7.58	_____
4. Kato, Naomi	8	8	8	8	8	_____	$12.28	_____
5. Freitag, Norman	6.5	4	9	7.5	7	_____	$10.42	_____
6. Scholl, Audrey	8	7.5	8	7.5	7.5	_____	$9.00	_____
7. Soto, Mario	6	0	8.5	6.5	8	_____	$9.62	_____
8. Johnson, Ray	5.5	8	7	7.5	7.5	_____	$8.04	_____
9. Thomas, Sandra	7	9	9	8.5	9.5	_____	$12.28	_____
10. Gregoir, Serge	4.5	8	8	7.5	7.5	_____	$8.56	_____

Name	Pieces Worked On						Piece Rate	Gross Income
	M	T	W	Th	F	Total		
11. Hurt, Nadia	106	87	98	102	86	_____	$0.70	_____
12. Maloney, Thomas	65	106	103	114	98	_____	$0.70	_____
13. Schwartz, Doris	79	68	72	73	65	_____	$0.94	_____
14. Tarantino, Dan	89	93	97	102	96	_____	$0.70	_____
15. Kavitsky, Irving	51	57	49	56	58	_____	$1.22	_____
16. MacLeod, Martha	52	65	68	69	61	_____	$0.94	_____
17. Flores, Rosa	61	62	64	59	56	_____	$1.22	_____
18. Czescik, Henry	106	112	120	106	109	_____	$0.70	_____
19. Cummings, Carla	32	37	36	31	32	_____	$1.96	_____
20. Kelly, Louise	71	63	76	72	61	_____	$0.94	_____

OPERATIONS WITH MONEY

Read each problem carefully. Think about which operations to use.

A fast-food restaurant lists the following prices;

Hamburger	$2.35	Ice cream	$1.45
Soft drink	0.68	Small fries	0.48
Shake	0.85	Large fries	0.76

EXAMPLES **If you buy a hamburger, soft drink, and small fries, how much change will you get from $5?**

(1) $2.35
     0.68
   + 0.48
     $3.51

(2) $5.00
    − 3.51
      $1.49   **Answer**

Find the cost of each of the following purchases.

1. 2 burgers and 2 shakes _____

2. 2 burgers, 2 soft drinks, and 1 large fries _____

3. 3 soft drinks and 1 large fries _____

4. 5 burgers, 4 ice creams, and 1 shake _____

5. 2 soft drinks and 2 large fries _____

6. 3 burgers and 3 shakes _____

For each of the following, find the amount of change you would get back from $10.

7. 3 burgers and 3 soft drinks _____

8. 3 ice creams and 2 soft drinks _____

9. 2 shakes and 1 large fries _____

10. 1 hamburger, 1 soft drink, 1 ice cream, and 1 small fries _____

11. 2 burgers, 2 shakes, 2 small fries _____

12. 5 shakes and 2 ice creams _____

OPERATIONS WITH MONEY

Read each problem carefully. Decide which operation to use. Write each item and amount. Then perform the operation called for.

The following prices were listed in a furniture store:

Easy chair	$219.00	Kitchen chair	$ 54.50
Kitchen table	76.50	Sofa	345.00
Desk	109.90	Bookcase	89.95

EXAMPLE **How much would you have to pay for a kitchen table and 4 chairs?**

1 kitchen chair	$ 54.50
	x 4
4 chairs	$218.00

4 chairs	$218.00
Table	+ 76.50
	$294.50

1. Find the cost of an easy chair, a desk, and a sofa.

Answer: _____

---

2. Find the cost of 2 kitchen chairs and 2 bookcases.

Answer: _____

3. Find the cost of 2 easy chairs, a sofa, and 6 kitchen chairs.

Answer: _____

---

4. If you bought an easy chair, a sofa, and a desk, how much would you have left from $1000?

Answer: _____

5. If you had $200, how much more would you need to buy a kitchen table and 6 chairs?

Answer: _____

48

DIVIDING MONEY

Place a decimal point in your answer over the decimal point in the dividend. Divide as you did for whole numbers.

EXAMPLES

$$\begin{array}{r} \$2.51 \\ 6\overline{)\$15.06} \\ \underline{12} \\ 3\,0 \\ \underline{3\,0} \\ 06 \\ \underline{06} \\ 0 \end{array}$$

$$\$8.24 \div 4 \rightarrow \begin{array}{r} \$2.06 \\ 4\overline{)\$8.24} \\ \underline{8} \\ 24 \\ \underline{24} \\ 0 \end{array}$$

Divide.

1. $2\overline{)\$18.30}$

2. $4\overline{)\$35.84}$

3. $3\overline{)\$261.06}$

4. $7\overline{)\$145.60}$

5. $5\overline{)\$331.45}$

6. $6\overline{)\$487.38}$

7. $9\overline{)\$276.75}$

8. $8\overline{)\$130.16}$

9. $7\overline{)\$277.90}$

10. $6\overline{)\$477.54}$

11. $9\overline{)\$268.83}$

12. $5\overline{)\$253.80}$

13. $\$50.49 \div 9 = $ _____

14. $\$26.43 \div 3 = $ _____

15. $\$67.95 \div 5 = $ _____

MULTIPLYING AND DIVIDING DECIMALS

When you multiply a decimal by 10, 100, or 1000, count the zeros in the multiplier and move the decimal point that many places to the right. When you divide a decimal by 10, 100, or 1000, count the zeros in the divisor and move the decimal point that many places to the left. Add zeros to the answer if necessary to fill in place values.

EXAMPLES   **29.2 x 100 = 2920**   **16.38 ÷ 1000 = .01638**

Multiply or divide.

1. 10 x 6.5 = _____

2. 8.152 x 1000 = _____

3. 100 x 47.3 = _____

4. 8.01 x 100 = _____

5. 0.5861 x 10 = _____

6. 94.3 x 1000 = _____

7. 1000 x 15.2 = _____

8. 6.1 x 1000 = _____

9. 10 x 2.4003 = _____

10. .197 x 10 = _____

11. 42.1 x 1000 = _____

12. 0.5983 x 100 = _____

13. 55.962 x 100 = _____

14. 1000 x 715.3 = _____

15. 10.246 x 100 = _____

16. 0.24639 x 10 = _____

17. 10 x 7.831 = _____

18. 8.402 x 100 = _____

19. 4500.1 x 100 = _____

20. 18.6 x 10 = _____

21. 9.512 x 100 = _____

22. 15.73 x 10 = _____

23. 16.4 x 100 = _____

24. 10 x .80 = _____

25. 10 x 9.1 = _____

26. 100 x 7.4 = _____

27. 7652.1 x 100 = _____

28. 8.02 x 10 = _____

29. 203.9 x 100 = _____

30. 360.5 x 100 = _____

31. 0.87 x 1000 = _____

32. 183.7 x 10 = _____

33. 6.8 x 100 = _____

34. 25.1 ÷ 10 = _____

35. 416.73 ÷ 100 = _____

36. 391.4 ÷ 100 = _____

37. 82.4 ÷ 100 = _____

38. 610.8 ÷ 100 = _____

39. 9138.2 ÷ 10 = _____

40. 26.05 ÷ 10 = _____

41. 3.39 ÷ 1000 = _____

42. 540.84 ÷ 100 = _____

43. 43.36 ÷ 10 = _____

44. 51.79 ÷ 100 = _____

45. 781.1 ÷ 1000 = _____

46. 125.39 ÷ 100 = _____

47. 9124.3 ÷ 100 = _____

48. 54 ÷ 1000 = _____

49. 64.3 ÷ 1000 = _____

50. 0.175 ÷ 10 = _____

51. 7.1 ÷ 100 = _____

52. 0.39 ÷ 100 = _____

53. 6.2 ÷ 1000 = _____

54. 1.29 ÷ 100 = _____

OPERATIONS WITH MONEY

EXAMPLE **If you earned $12 working for 5 days after school, find the average amount earned each day.**

$$\begin{array}{r} \$\ 2.40 \\ 5\overline{)\$12.00} \end{array}$$

---

1. Find the amount earned in 9 hours at $6.50 per hour.

Answer: _____

2. Find the amount earned in 6 weeks at $37 per week.

Answer: _____

---

3. Find the amount earned in 7 months at $157 per month.

Answer: _____

4. Find the amount earned per day at $364 for 8 days.

Answer: _____

---

5. Find the amount earned per month at $762 for 6 months.

Answer: _____

6. Find the amount earned per day at $1235 for 25 days.

Answer: _____

---

7. Find the amount left from $100 if you spent $6 per day for 7 days.

Answer: _____

8. Find the amount earned in a year at $1512 per month.

Answer: _____

---

9. Find the amount left from $1000 if you spent $28 per week for 34 weeks.

Answer: _____

10. Find the amount left from $5000 if you spent $326 per month for 7 months.

Answer: _____

OPERATIONS WITH MONEY

Read each problem carefully. Select the right operation.

EXAMPLE **If you save $24 a week, how long will it take to save $360?**

Since $24 x number of weeks = $360, you must divide to get the answer.

```
        15 weeks
   24)360
      24
     120
     120
       0
```

1. If you saved $6 a week, how long would it take you to save $102?

Answer: _____

---

2. If a car costs $4428, and you pay $108 a month, how long will it take you to pay for it?

Answer: _____

3. If you spent $9 each week, how much would you spend in a year?

Answer: _____

---

4. If you had $90 and saved another $26 a month for 6 months, how much would you have at the end of the 6 months?

Answer: _____

5. If you saved $16 a week, how long would you have to save to buy a television set costing $288?

Answer: _____

---

6. If you paid $0.75 to cross the Oldport Bridge, how much would you pay to cross the bridge 230 times?

Answer: _____

7. If a car costs $7499.52, and you pay for it over 48 months, how much must you pay per month?

Answer: _____

52

ROUNDING MONEY

Round as you do with whole numbers.

EXAMPLES **Round to the nearest hundred dollars.**

$725 7 is the rounding number.
      2 is to the right of 7.
      2 is less than 5.
$700 Answer

$899 8 is the rounding number.
      9 is to the right of 8.
      9 is greater than 5.
$900 Answer

Round to the nearest hundred dollars.

1. $614 _____

2. $487 _____

3. $295 _____

4. $954 _____

5. $109 _____

6. $260 _____

7. $459 _____

8. $399 _____

9. $995 _____

10. $555 _____

11. $108 _____

12. $404 _____

13. $670 _____

14. $201 _____

15. $959 _____

16. $602 _____

17. $378 _____

18. $615 _____

19. $487 _____

20. $3752 _____

21. $7841 _____

22. $2309 _____

23. $3862 _____

24. $2719 _____

25. $1198 _____

26. $2451 _____

27. $284 _____

28. $833 _____

29. $3796 _____

30. $725 _____

31. $6258 _____

32. $9016 _____

33. $253 _____

34. $71,305 _____

35. $327.56 _____

36. $914.80 _____

37. $155.68 _____

38. $820.99 _____

39. $495.21 _____

40. $9302.63 _____

41. $627.61 _____

42. $194.20 _____

43. $683.56 _____

44. $2390.28 _____

45. $597.30 _____

46. $639.19 _____

47. $871.28 _____

48. $3964.82 _____

49. $364.15 _____

50. $583.86 _____

51. $769.37 _____

52. $741.19 _____

53. $8592.04 _____

54. $946.19 _____

55. $557.92 _____

56. $4520.46 _____

57. $715.83 _____

58. $278.23 _____

59. $499.00 _____

60. $998.05 _____

ROUNDING MONEY

Round to the nearest dollar.

$995   9 is the rounding number. $6.02 6 is the rounding number.
            9 is to the right of 9.   0 is to the right of 6.
            9 is greater than 5.   0 is less than 5.
$10.00 Answer $6.00 Answer

Round to the nearest dollar.

1. $1.15 _____   2. $8.85 _____   3. $7.75 _____   4. $4.50 _____

5. $0.95 _____   6. $19.95 _____   7. $9.70 _____   8. $6.05 _____

9. $0.98 _____   10. $17.20 _____   11. $9.12 _____   12. $1.90 _____

13. $29.95 _____   14. $0.15 _____   15. $99.95 _____   16. $38.16 _____

17. $7.00 _____   18. $2.43 _____   19. $87.59 _____   20. $14.49 _____

21. $62.09 _____   22. $8.76 _____   23. $76.72 _____   24. $19.63 _____

25. $7.39 _____   26. $0.62 _____   27. $7.77 _____   28. $38.83 _____

Add. Then round to the nearest dollar.

29.	30.	31.
$ 98.14	$324.08	$ 2.17
2.79	.03	47.88
109.65	46.85	109.40
.43	196.48	81.93
+ 2.76	+ 26.14	+ 51.15

32.	33.	34.
$ 73.19	$ 64.38	$119.21
31.88	1.96	6.09
9.16	173.75	317.90
113.53	292.00	57.57
+ .92	+ 6.17	+ 41.18

35.	36.	37.
$ 64.17	$716.39	$ 68.34
.68	128.05	265.19
35.19	16.97	35.56
186.62	8.22	7.04
+ 70.09	+ 636.77	+ 108.83

ROUNDING MONEY

Round to the nearest cent.

$2.3862 ⟶ $2.39   **Round up since 6 is greater than 5.**

$4.14
x 0.8
$3.312 ⟶ $3.31   **Round down since 2 is less than 5.**

Round to the nearest cent.

1. $2.614 _____

2. $59.183 _____

3. $16.918 _____

4. $4.1684 _____

5. $0.1829 _____

6. $4,6075 _____

7. $30.1991 _____

8. $5.0253 _____

9. $94.2571 _____

10. $8.23542 _____

11. $0.95678 _____

12. $46.3020 _____

Multiply. Then round to the nearest cent.

13. $36.25
    x   0.3

14. $46.39
    x   0.8

15. $96.45
    x   0.9

16. $225.63
    x   0.5

17. $963.55
    x   0.7

18. $200.63
    x   0.4

19. $37.86
    x 0.35

20. $83.19
    x 0.63

21. $44.33
    x 0.69

22. $601.77
    x   0.74

23. $436.35
    x   0.09

24. $692.47
    x   0.75

25. $27.39
    x 0.063

26. $84.33
    x 0.176

27. $91.54
    x 0.126

OPERATIONS USING ROUNDING

The answer to each problem is an estimate. Each estimate is a round number, such as 50, 600, or 1000. You can find the answer by rounding before you do the operation called for.

EXAMPLE **If $5021 is divided among 5 people, about how much money will each person receive?**

$5021 is about $5000.
$5000 ÷ 5 = $1000
Each person will receive about $1000.

1. If $198 is divided between 2 people, about how much will each receive?

Answer: _____

2. If a car travels 201 miles in 4 hours, about how far does it go in 1 hour?

Answer: _____

3. If 1 pound of meat costs $798, about how much will 3 pounds cost?

Answer: _____

4. If 1 baseball costs $5.98, about how much will 5 baseballs cost?

Answer: _____

5. If a plane travels 800 miles in 58 minutes, about how far will it travel in 2 hours?

Answer: _____

6. If 3 CDs cost $36.05, about how much will 1 CD cost?

Answer: _____

7. If you earn $7.82 an hour, about how much will you earn in 3 hours?

Answer: _____

8. If a car travels 40 miles in 61 minutes, about how long will it take to travel 80 miles?

Answer: _____

9. If 1 gallon of paint will cover 202 square feet, about how much paint will it take to cover 400 square feet?

Answer: _____

ROUNDING MONEY

EXAMPLE **If you borrow money from a bank, you pay back part of the loan and part of the total interest each month. Suppose you borrow $1500 for 36 months. The total interest is $432.56. Find the approximate monthly payment.**

$1500.00 Loan
+ 432.56 Interest
$1932.56 Total

$53.682 ───────▶ $53.68 Answer
36)$1932.560

Amount Borrowed	Interest for 24 months	Interest for 36 months	Interest for 48 months
$ 500	$ 91.23	$144.19	$ 202.56
1000	182.46	288.38	405.11
1500	273.69	432.56	607.67
2000	364.92	576.75	810.22
2500	456.15	720.94	1012.78
3000	547.38	865.13	1215.33

Use the table above to find the monthly payment for each loan.

1. $1000 for 24 months _____

2. $2500 for 24 months _____

3. $3000 for 48 months _____

4. $1500 for 48 months _____

5. $2500 for 36 months _____

6. $500 for 36 months _____

7. $2500 for 48 months _____

8. $3000 for 36 months _____

9. $1500 for 24 months _____

10. $2000 for 48 months _____

ROUNDING MONEY

Pencils are on sale at 7 for $1. How much will 1 pencil cost?

$$\frac{0.142}{7)\overline{\$1.000}} \longrightarrow \$0.15 \quad \text{Store owners always round up.}$$

In each problem, find the cost of one item by dividing and rounding up.

1. 2 cans of tomatoes for $1.79	**2.** 5 pounds of potatoes for $1.69	**3.** 3 bars of soap for $0.71
Answer: _____	Answer: _____	Answer: _____
4. 2 CDs for $24.95	**5.** 7 ball-point pens for $1.99	**6.** 7 oranges for $3
Answer: _____	Answer: _____	Answer: _____

For each item, decide which container is the better buy. Circle the answer.

EXAMPLE
Vinegar
8 ounces for $0.79
20 ounces for $1.79
 	
$\dfrac{0.098}{8)\overline{\$0.790}} \longrightarrow \$0.10$
 	
$\dfrac{0.089}{20)\overline{\$1.790}} \longrightarrow \$0.09$
	7. Applesauce
	    16 ounces for $1.39
	    24 ounces for $2.39
		8. Toothpaste
		    4 ounces for $1.59
		    7 ounces for $2.49

9. Ketchup
    32 ounces for $2.89
    12 ounces for $1.29
	10. Salt
	    12 ounces for $0.89
	    18 ounces for $1.59
		11. Spaghetti
		    2 pounds for $1.49
		    3 pounds for $2.20

SOLVING EQUATIONS BY ADDITION AND SUBTRACTION

You may add or subtract the same number on both sides of an equation.

EXAMPLES

$$x + 5 = 11$$
$$x + 5 - 5 = 11 - 5$$
$$x = 6$$

$$a - 7 = 3.5$$
$$a - 7 + 7 = 3.5 + 7$$
$$a = 10.5$$

Solve each equation.

1. $x + 6 = 10$

2. $y - 4 = 6$

3. $s + 21 = 100$

4. $a - 51 = 41$

5. $b - 3 = 2$

6. $c + 77 = 326$

7. $d + 3.4 = 11$

8. $x - 1.1 = 4.5$

9. $y - 8.3 = 16.2$

10. $a - 150 = 25$

11. $b + 3.05 = 7.19$

12. $c - 2.3 = 4.15$

13. $x - 32 = 325$

14. $1.2 + y = 2.0$

15. $2 + a = 7$

16. $b + 5 = 21$

17. $c - 3.2 = 5$

18. $s + 100 = 101$

SOLVING EQUATIONS BY ADDITION AND SUBTRACTION

REMEMBER: You can use the same rules for adding or subtracting when the variable (letter) is on the right side of the equation.

EXAMPLES
$$5 = 3 + x$$
$$5 - 3 = 3 - 3 + x$$
$$2 = x$$

$$7 = a - 3$$
$$7 + 3 = a - 3 + 3$$
$$10 = a$$

Solve each equation.

1. $7 = a + 3$

2. $40 = b - 10$

3. $9 = c + 3$

4. $2.5 = x + 1.5$

5. $6 = a - 3.5$

6. $5 = b - 2$

7. $x - 4 = 19$

8. $s + 15 = 25$

9. $17 = y - 3.4$

10. $7.9 = a - 2.1$

11. $x - 3.4 = 12$

12. $15.5 = a + 2.2$

13. $5 + 7 = x - 1$

14. $78 = y + 45$

15. $100 = a + 45$

16. $18 = 10 + x$

17. $2.1 = 0.5 + y$

18. $a - 4 = 3.5$

SOLVING MULTI-STEP PROBLEMS

REMEMBER: Some problems take more than one calculation to solve. First, think about what you can calculate to help solve the problem. Then use that answer to solve the problem.

EXAMPLE **The following prices were listed at the Seaside Cafe:**

Sandwich	$3.75	Lemonade	$1.50
Hamburger	$2.95	Iced Tea	$1.75
Salad	$2.49	Soda	$1.85
Yogurt	$1.75	Ice Cream	$2.25

Ms. Williams bought a sandwich, yogurt, and iced tea for lunch. She paid with a ten-dollar bill. How much change should she receive?

Think: Which is easier to find first:

  • **the cost of the lunch?**   • **the change she received?**

The cost of her lunch is easier to find first.

$$\begin{array}{ll} (1)\ \ \$3.75 & (2)\ \ \$10.00 \\ \ \ \ \ \ \ 1.75 & \ \ \ \ \ -7.25 \\ \ \ \ +1.75 & \ \ \ \ \$\ 2.75 \\ \ \ \$7.25 & \end{array}$$

The cost of the lunch is $7.25. Ms. Williams' change is $2.75.

---

1. Harvey had a hamburger, a lemonade, and ice cream for lunch. Mary had a salad, a soda, and ice cream. Whose lunch cost more?

   Answer: _____

2. Mr. Valez bought a sandwich, a lemonade, and ice cream for each of his 3 children. How many ten-dollar bills will he need to pay for lunch?

   Answer: _____

---

3. A cheeseburger costs $0.25 more than a hamburger. How much do 3 cheeseburgers and 3 iced teas cost?

   Answer: _____

4. The special price for 10 ice creams is $20.00. How much of a savings is that over the regular price?

   Answer: _____

The table shows the fares for railroad travel between Los Santos and nearby cities.

RAILROAD TRAVEL FROM LOS SANTOS
(Children's tickets are half price.)

City	One-way Ticket	Round Trip Ticket
Freeman	$4.80	$8.20
Gainesville	5.50	9.50
Hunter City	6.30	10.40

5. Mrs. Banta lives in Los Santos. She buys a round trip ticket to Gainesville for herself and her two children. How much do the 3 tickets cost?

Answer: _____

6. Isabel buys a one-way ticket from Freeman to Los Santos. Later, she buys a one-way ticket back to Freeman. How much money would she have saved by buying a round trip ticket instead?

Answer: _____

7. Ms. Moss, her father, and her child are traveling from Los Santos to Gainesville. Will $15 be enough for all 3 tickets?

Answer: _____

8. Gerald commutes to work from Freeman to Los Santos. He can buy ten round trip tickets for $75.00. How much does he save on those ten rides?

Answer: _____

9. Mr. Clark is taking his grandson from Los Santos to Hunter City to visit the museum. For the two round trip tickets, how much change will he receive from $20?

Answer: _____

10. Vinny lives in Hunter City. He is going to Los Santos for the day and will then spend the night in Freeman. He can use 2 one-way tickets or a round trip ticket between Hunter City and Los Santos. Which costs less?

Answer: _____

CHAPTER 2

POSTTEST

Add.

1. $2.20
   + 3.38

2. $93.14
   16.49
   + 58.80

3. 49.17
   2.6825
   +27.3

4. 7 + 5.21 + .608 = _____

Subtract.

5. $47.69
   − 3.05

6. $72.43
   − 9.86

7. 6.13
   − 1.092

8. 8.57 − 3.6 = _____

Multiply.

9. $16.28
   x   7

10. $63.17
    x   85

11. 34.9
    x 5.4

12. 8.35 x 7.2 = _____

Divide.

13. 6)27.6

14. 15)$72.45

15. 3.1)9.61

16. 6.40 ÷ 0.8 = _____

Round off $673.524 to the nearest:

17. Dollar _____

18. Hundred dollars _____

19. Cent _____

Work out the following exercises.

20. Add $61.53, $90.26, and $63. _____

21. Subtract $9.05 from $30. _____

22. Multiply twelve dollars five cents by eighty-seven. _____

Solve each equation.

23. $x + 12 = 41$

24. $1.6 = a + 0.6$

25. A hat costs $5.40 and a belt costs $7.95. Find the total cost of both items.

Answer: _____

26. If you bought a book for $7.95, what would be your change from $10?

Answer: _____

27. A chair costs $32.70. Find the cost of 5 chairs.

Answer: _____

28. If a hot dog costs $1.49 and juice costs $0.39, find the cost of 3 hot dogs and 2 juices.

Answer: _____

29. If you earned $455 in 7 days, how much would that be for each day?

Answer: _____

30. If you paid $53.55 for 9 tickets to a ball game, what was the cost of each ticket?

Answer: _____

31. If you saved $30 per week, how long would it take you to save $210?

Answer: _____

32. If you earned $45 per week, how much would you earn in 12 weeks?

Answer: _____

33. Circle the largest of the three numbers.   0.605   0.65   0.6

64

PRETEST

1. Write the fraction that matches the shaded part.

_____

2. Shade the figure to match the given fraction.

$\frac{5}{6}$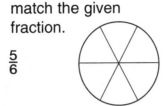

3. Draw a figure and shade it to match the fraction.

$\frac{3}{4}$

4. Reduce to lowest terms.

$\frac{6}{16} =$ _____

5. Convert to a fraction with the denominator 12.

$\frac{3}{4} = \overline{12}$

6. Convert to a mixed number.

$\frac{25}{4} =$ _____

7. Convert to an improper fraction.

$3\frac{5}{6} =$ _____

8.
$\begin{array}{r} \frac{1}{8} \\ + \frac{3}{8} \\ \hline \end{array}$

9.
$\begin{array}{r} \frac{1}{2} \\ + \frac{3}{8} \\ \hline \end{array}$

10.
$\begin{array}{r} 1\frac{1}{2} \\ + 2\frac{2}{3} \\ \hline \end{array}$

11. $\frac{7}{20} - \frac{1}{20} =$ _____

12. $\frac{3}{5} - \frac{1}{4} =$ _____

13.
$\begin{array}{r} 8\frac{5}{8} \\ - 4\frac{1}{2} \\ \hline \end{array}$

14.
$\begin{array}{r} 4\frac{1}{4} \\ - 2\frac{3}{4} \\ \hline \end{array}$

15. $\frac{2}{3} \times \frac{4}{5} =$ _____

16. $6 \times \frac{2}{3} =$ _____

17. $1\frac{3}{4} \times \frac{1}{2} =$ _____

18. $\frac{7}{16} \div \frac{1}{8} =$ _____

19. $5\frac{1}{3} \div \frac{2}{3} =$ _____

20. $\frac{5}{6} \div 6 =$ _____

21. $\frac{1}{2} \div 3\frac{1}{3} =$ _____

22. If you started with $25 and spent two-fifths of it, how much money did you spend?

Answer: _____

23. If you waited $\frac{2}{3}$ hour for someone, how many minutes did you wait?

Answer: _____

24. If you spent one-third of your money for a movie and one-fourth for lunch, what part did you spend in all?

Answer: _____

25. A worker earned $960. She spent one-tenth of it. How much money did she have left?

Answer: _____

26. If there were 15 people in a class and 3 of them came late, what part of the class came late?

Answer: _____

27. If 90 people were at a game and one-sixth bought programs for $1.50 each, how much did they spend on programs all together?

Answer: _____

28. If you had $140 and lost one-fourth of it, how much money would you have left?

Answer: _____

29. How many hours are there in $\frac{3}{4}$ day?

Answer: _____

30. Convert $\frac{3}{5}$ to a decimal.

Answer: _____

31. Convert .06 to a fraction.

Answer: _____

32. Solve for x. $8x = 20$

Answer: _____

33. Solve for b. $\frac{b}{2} = 1\frac{1}{2}$

Answer: _____

THE MEANING OF A FRACTION

A fraction has a top number and a bottom number. The top number is the numerator. The bottom number is the denominator.

EXAMPLES

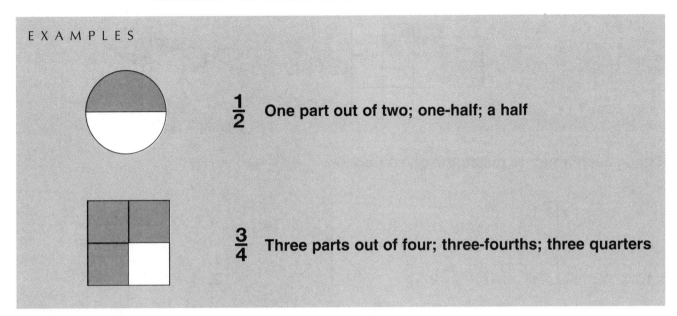

$\frac{1}{2}$ **One part out of two; one-half; a half**

$\frac{3}{4}$ **Three parts out of four; three-fourths; three quarters**

For each figure, write the fraction that matches the shaded part.

1. _____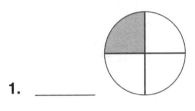

2. _____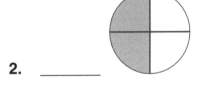

3. _____

4. _____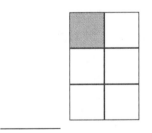

5. _____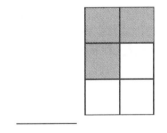

6. _____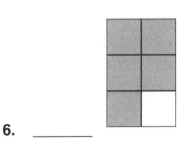

7. _____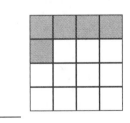

8. _____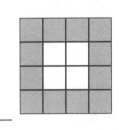

9. _____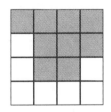

REPRESENTING FRACTIONS

The denominator, the bottom number, tells you the total number of parts. The numerator, the top number, tells you how many parts of the total.

EXAMPLES

$\frac{2}{9}$ 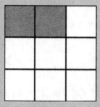 $\frac{1}{3}$

Shade each figure to match the given fraction.

1. $\frac{1}{6}$ 2. $\frac{5}{8}$ 3. $\frac{3}{4}$

4. $\frac{4}{9}$ 5. $\frac{7}{12}$    6. $\frac{4}{6}$

Divide each figure into the number of parts shown in the denominator of the given fraction. Then shade to represent the fraction.

7. $\frac{3}{8}$ 8. $\frac{5}{9}$ 9. $\frac{5}{6}$

10. $\frac{3}{12}$ 11. $\frac{2}{4}$ 12. $\frac{2}{3}$

EQUIVALENT FRACTIONS

There are many ways of writing the same fraction.

EXAMPLE **Convert $\frac{2}{6}$ to a fraction with the denominator 12.**

First write $\frac{2}{6} = \frac{}{12}$. **Multiply the numerator and the denominator by the same number.** $\frac{2 \times 2}{2 \times 6} = \frac{4}{12}$ **Answer**

Convert each of the following fractions to an equivalent fraction with the denominator 12.

1. $\frac{1}{4} =$ _____ 2. $\frac{1}{3} =$ _____ 3. $\frac{1}{6} =$ _____ 4. $\frac{2}{3} =$ _____ 5. $\frac{3}{4} =$ _____

6. $\frac{1}{2} =$ _____ 7. $\frac{5}{6} =$ _____ 8. $\frac{4}{6} =$ _____ 9. $\frac{2}{4} =$ _____ 10. $\frac{3}{6} =$ _____

Convert each of the following fractions to an equivalent fraction with the denominator 20.

11. $\frac{1}{2} =$ _____ 12. $\frac{1}{5} =$ _____ 13. $\frac{3}{4} =$ _____ 14. $\frac{2}{10} =$ _____ 15. $\frac{8}{10} =$ _____

16. $\frac{7}{10} =$ _____ 17. $\frac{4}{5} =$ _____ 18. $\frac{9}{10} =$ _____ 19. $\frac{2}{4} =$ _____ 20. $\frac{6}{10} =$ _____

21. $\frac{1}{4} =$ _____ 22. $\frac{3}{5} =$ _____ 23. $\frac{1}{10} =$ _____ 24. $\frac{2}{5} =$ _____ 25. $\frac{4}{10} =$ _____

Convert to an equivalent fraction with the denominator 100.

26. $\frac{1}{5} =$ _____ 27. $\frac{3}{20} =$ _____ 28. $\frac{4}{25} =$ _____ 29. $\frac{6}{10} =$ _____ 30. $\frac{9}{50} =$ _____

31. $\frac{19}{20} =$ _____ 32. $\frac{3}{4} =$ _____ 33. $\frac{8}{25} =$ _____ 34. $\frac{9}{10} =$ _____ 35. $\frac{3}{5} =$ _____

36. $\frac{1}{4} =$ _____ 37. $\frac{7}{20} =$ _____ 38. $\frac{17}{50} =$ _____ 39. $\frac{21}{25} =$ _____ 40. $\frac{3}{10} =$ _____

REDUCING FRACTIONS

A fraction is in lowest terms when it cannot be reduced.

EXAMPLES Reduce $\frac{42}{56}$ to lowest terms. $\frac{42}{56} =$

Both 42 and 56 are divisible by 7. $\frac{42 \div 7}{56 \div 7} = \frac{6}{8}$

Both 6 and 8 are divisible by 2. $\frac{6 \div 2}{8 \div 2} = \frac{3}{4}$

Reduce each fraction to lowest terms.

1. $\frac{5}{10} = $ _____

2. $\frac{9}{24} = $ _____

3. $\frac{28}{35} = $ _____

4. $\frac{16}{20} = $ _____

5. $\frac{21}{24} = $ _____

6. $\frac{6}{8} = $ _____

7. $\frac{45}{54} = $ _____

8. $\frac{40}{50} = $ _____

9. $\frac{4}{8} = $ _____

10. $\frac{10}{15} = $ _____

11. $\frac{12}{20} = $ _____

12. $\frac{40}{60} = $ _____

13. $\frac{7}{28} = $ _____

14. $\frac{10}{25} = $ _____

15. $\frac{36}{40} = $ _____

16. $\frac{5}{20} = $ _____

17. $\frac{11}{99} = $ _____

18. $\frac{27}{36} = $ _____

19. $\frac{10}{70} = $ _____

20. $\frac{15}{60} = $ _____

21. $\frac{36}{60} = $ _____

22. $\frac{18}{36} = $ _____

23. $\frac{42}{60} = $ _____

24. $\frac{12}{16} = $ _____

25. Write 15 minutes as a fractional part of an hour and reduce to lowest terms.

_____

26. Write 3 hours as a fractional part of a day and reduce to lowest terms.

_____

27. Write 2 months as a fractional part of a year and reduce to lowest terms.

_____

28. Write 10 years as a fractional part of a century and reduce to lowest terms.

_____

CONVERTING MIXED NUMBERS TO IMPROPER FRACTIONS

Convert each mixed number to an improper fraction.

1. $3\frac{1}{4} =$ _____

2. $1\frac{1}{2} =$ _____

3. $4\frac{1}{5} =$ _____

4. $7\frac{1}{3} =$ _____

5. $5\frac{2}{4} =$ _____

6. $4\frac{1}{2} =$ _____

7. $3\frac{1}{9} =$ _____

8. $7\frac{2}{9} =$ _____

9. $2\frac{1}{4} =$ _____

10. $4\frac{5}{7} =$ _____

11. $6\frac{3}{5} =$ _____

12. $3\frac{5}{9} =$ _____

13. $7\frac{3}{5} =$ _____

14. $8\frac{5}{7} =$ _____

15. $12\frac{1}{4} =$ _____

16. $11\frac{5}{7} =$ _____

17. $5\frac{1}{6} =$ _____

18. $11\frac{2}{3} =$ _____

19. $6\frac{1}{3} =$ _____

20. $4\frac{4}{7} =$ _____

21. $13\frac{1}{2} =$ _____

22. $7\frac{2}{5} =$ _____

23. $3\frac{3}{12} =$ _____

24. $10\frac{1}{4} =$ _____

25. $5\frac{3}{7} =$ _____

26. $10\frac{2}{9} =$ _____

27. $9\frac{3}{10} =$ _____

28. $8\frac{5}{12} =$ _____

29. $4\frac{3}{6} =$ _____

30. $8\frac{2}{5} =$ _____

31. $7\frac{5}{6} =$ _____

32. $5\frac{2}{7} =$ _____

33. $10\frac{2}{11} =$ _____

34. $2\frac{11}{13} =$ _____

35. $3\frac{7}{10} =$ _____

36. $5\frac{5}{8} =$ _____

37. $5\frac{7}{10} =$ _____

38. $7\frac{7}{8} =$ _____

39. $5\frac{10}{13} =$ _____

40. $6\frac{1}{8} =$ _____

41. $4\frac{11}{15} =$ _____

42. $7\frac{13}{16} =$ _____

43. $9\frac{3}{14} =$ _____

44. $10\frac{7}{11} =$ _____

RENAMING IMPROPER FRACTIONS

REMEMBER: Divide the numerator by the denominator to find the whole part. Any remainder is the fractional part.

EXAMPLES

$\frac{5}{3} \longrightarrow 3\overline{)5}$ $\frac{5}{3} = 1\frac{2}{3}$ $\frac{15}{5} \longrightarrow 5\overline{)15}$ $\frac{15}{5} = 3$

$\begin{array}{r} 1 \\ 3\overline{)5} \\ 3 \\ \hline 2 \end{array}$ $\begin{array}{r} 3 \\ 5\overline{)15} \\ 15 \\ \hline 0 \end{array}$

Convert each improper fraction to a mixed number or a whole number.

1. $\frac{3}{2} = 2\overline{)3}$ $\begin{array}{r} 1\frac{1}{2} \\ 2\overline{)3} \\ 2 \\ \hline 1 \end{array}$ 2. $\frac{7}{5} = \overline{)}$ 3. $\frac{4}{4} = \overline{)}$ 4. $\frac{5}{3} = \overline{)}$ 5. $\frac{15}{2} = \overline{)}$

6. $\frac{7}{2} = \overline{)}$ 7. $\frac{6}{3} = \overline{)}$ 8. $\frac{9}{4} = \overline{)}$ 9. $\frac{21}{5} = \overline{)}$ 10. $\frac{15}{10} = \overline{)}$

11. $\frac{9}{6} = \overline{)}$ 12. $\frac{12}{10} = \overline{)}$ 13. $\frac{15}{3} = \overline{)}$ 14. $\frac{20}{12} = \overline{)}$ 15. $\frac{21}{20} = \overline{)}$

16. $\frac{7}{4} = \overline{)}$ 17. $\frac{13}{8} = \overline{)}$ 18. $\frac{12}{8} = \overline{)}$ 19. $\frac{8}{5} = \overline{)}$ 20. $\frac{40}{30} = \overline{)}$

21. $\frac{8}{7} = \overline{)}$ 22. $\frac{11}{2} = \overline{)}$ 23. $\frac{11}{10} = \overline{)}$ 24. $\frac{12}{12} = \overline{)}$ 25. $\frac{11}{5} = \overline{)}$

26. $\frac{8}{3} = \overline{)}$ 27. $\frac{10}{10} = \overline{)}$ 28. $\frac{10}{4} = \overline{)}$ 29. $\frac{18}{6} = \overline{)}$ 30. $\frac{21}{20} = \overline{)}$

31. $\frac{9}{2} = \overline{)}$ 32. $\frac{31}{30} = \overline{)}$ 33. $\frac{13}{6} = \overline{)}$ 34. $\frac{60}{3} = \overline{)}$ 35. $\frac{90}{2} = \overline{)}$

WRITING FRACTIONS AS DECIMALS

To convert a fraction to a decimal, divide the numerator by the denominator.

EXAMPLES **Write $\frac{5}{8}$ as a decimal.**

$$
\begin{array}{r}
.625 \\
8\overline{)5.000} \\
\underline{4\,8} \\
20 \\
\underline{16} \\
40 \\
\underline{40} \\
0
\end{array}
$$

Write $\frac{41}{40}$ as a decimal.

$$
\begin{array}{r}
1.025 \\
40\overline{)41.000} \\
\underline{40} \\
1\,00 \\
\underline{80} \\
200 \\
\underline{200} \\
0
\end{array}
$$

Convert each fraction to a decimal. Keep dividing until the remainder is zero.

1. $\frac{3}{4} =$

2. $\frac{1}{2} =$

3. $\frac{2}{5} =$

4. $\frac{11}{8} =$

5. $\frac{4}{10} =$

6. $\frac{1}{8} =$

7. $\frac{41}{20} =$

8. $\frac{3}{16} =$

9. $\frac{27}{12} =$

10. $\frac{19}{40} =$

11. $\frac{7}{8} =$

12. $\frac{6}{25} =$

13. $\frac{9}{50} =$

14. $\frac{27}{20} =$

15. $\frac{5}{16} =$

WRITING FRACTIONS AS DECIMALS

Divide the numerator by the denominator.

EXAMPLES **Convert $\frac{2}{3}$ to a decimal. Round off to the nearest thousandth.**

$$.6666 \longrightarrow .667$$
$$3\overline{)2.0000}$$
$$\underline{1\ 8}$$
$$20$$
$$\underline{18}$$
$$20$$
$$\underline{18}$$
$$20$$

Convert $\frac{1}{7}$ to a decimal. Round off to the nearest thousandth.

$$.1428 \longrightarrow .143$$
$$7\overline{)1.0000}$$
$$\underline{7}$$
$$30$$
$$\underline{28}$$
$$20$$
$$\underline{14}$$
$$60$$

Convert each fraction to a decimal. Round to the nearest thousandth.

1. $\frac{1}{3} =$ _____

2. $\frac{2}{7} =$ _____

3. $\frac{1}{6} =$ _____

4. $\frac{7}{9} =$ _____

5. $\frac{4}{7} =$ _____

6. $\frac{2}{9} =$ _____

7. $\frac{5}{11} =$ _____

8. $\frac{7}{12} =$ _____

9. $\frac{2}{14} =$ _____

10. $\frac{9}{13} =$ _____

11. $\frac{5}{6} =$ _____

12. $\frac{7}{15} =$ _____

13. $\frac{5}{18} =$ _____

14. $\frac{10}{24} =$ _____

15. $\frac{6}{21} =$ _____

WRITING DECIMALS AS FRACTIONS

The denominator of the fraction will have as many zeros as there are decimal places in the original decimal number.

EXAMPLES .48   The last place to the right is hundredths. $.48 = \frac{48}{100} = \frac{24}{50} = \frac{12}{25}$
So .48 = 48 hundredths.

.008   The last place to the right is thousandths. $.008 = \frac{8}{1000} = \frac{1}{125}$
So .008 = 8 thousandths.

Convert each decimal to a fraction in lowest terms.

1. .8 = _____   **2.** .4 = _____   **3.** .9 = _____

4. .25 = _____   **5.** .55 = _____   **6.** .48 = _____

7. .63 = _____   **8.** .97 = _____   **9.** .96 = _____

10. .88 = _____   **11.** .05 = _____   **12.** .36 = _____

13. .075 = _____   **14.** .600 = _____   **15.** .350 = _____

16. .009 = _____   **17.** . 084 = _____   **18.** .305 = _____

19. 3.5 = _____   **20.** 6.8 = _____   **21.** 4.90 = _____

22. 1.56 = _____   **23.** 3.005 = _____   **24.** 9.250 = _____

25. .004 = _____   **26.** .38 = _____   **27.** 2.7 = _____

28. 6.20 = _____   **29.** 4.06 = _____   **30.** .70 = _____

31. Forty-five cents is what fractional part of a dollar? _____

32. Six cents is what fractional part of a dollar? _____

MULTIPLYING FRACTIONS

Multiply numerator by numerator and denominator by denominator.

EXAMPLES $\frac{2}{5} \times \frac{1}{3} = \frac{2}{15}$ $\frac{1}{5}$ of $\frac{3}{4} = \frac{3}{20}$ $\frac{1}{2} \times \frac{5}{7} \times \frac{1}{3} = \frac{5}{42}$

Multiply.

1. $\frac{2}{3} \times \frac{1}{3} =$ _____

2. $\frac{1}{5} \times \frac{1}{6} =$ _____

3. $\frac{3}{5} \times \frac{1}{4} =$ _____

4. $\frac{1}{2} \times \frac{5}{6} =$ _____

5. $\frac{2}{3} \times \frac{2}{9} =$ _____

6. $\frac{3}{10} \times \frac{3}{4} =$ _____

7. $\frac{3}{7} \times \frac{4}{5} =$ _____

8. $\frac{3}{5} \times \frac{2}{5} =$ _____

9. $\frac{2}{3} \times \frac{4}{5} =$ _____

10. $\frac{3}{4} \times \frac{1}{8} =$ _____

11. $\frac{2}{5} \times \frac{4}{7} =$ _____

12. $\frac{5}{6} \times \frac{1}{7} =$ _____

13. $\frac{5}{7}$ of $\frac{5}{8} =$ _____

14. $\frac{1}{2}$ of $\frac{1}{5} =$ _____

15. $\frac{3}{10}$ of $\frac{1}{5} =$ _____

16. $\frac{7}{10}$ of $\frac{11}{12} =$ _____

17. $\frac{3}{5}$ of $\frac{9}{10} =$ _____

18. $\frac{2}{3}$ of $\frac{8}{9} =$ _____

19. $\frac{3}{4} \times \frac{1}{2} \times \frac{5}{8} =$ _____

20. $\frac{2}{3} \times \frac{1}{5} \times \frac{4}{7} =$ _____

21. $\frac{7}{8} \times \frac{5}{9} \times \frac{1}{4} =$ _____

22. $\frac{2}{8} \times \frac{1}{3} \times \frac{11}{12} =$ _____

23. $\frac{4}{5} \times \frac{8}{9} \times \frac{2}{3} =$ _____

24. $\frac{4}{7} \times \frac{1}{3} \times \frac{5}{9} =$ _____

25. $\frac{13}{17} \times \frac{5}{19} =$ _____

26. $\frac{4}{15} \times \frac{16}{17} =$ _____

27. $\frac{9}{17} \times \frac{3}{7} =$ _____

28. Marty walked halfway to school and rode the bus the rest of the way. It is $\frac{7}{10}$ mile from home to school. How far did Marty walk?

29. It took Jose $\frac{1}{3}$ hour to do his home work. Marie did hers in half the time. What fraction of an hour did she spend on it?

Answer: _____

Answer: _____

76

MULTIPLYING FRACTIONS

EXAMPLE Multiply. Then reduce. $\frac{3}{8} \times \frac{4}{5} = \frac{12}{40} = \frac{3}{10}$

You may want to use a shortcut. When possible, divide both a numerator and a denominator by the same number.

$\frac{3}{\underset{2}{8}} \times \frac{\overset{1}{4}}{5} = \frac{3}{10}$

Multiply. Then reduce each fraction to lowest terms.

1. $\frac{3}{5} \times \frac{2}{3} = $ _____

2. $\frac{4}{5} \times \frac{7}{8} = $ _____

3. $\frac{2}{3} \times \frac{6}{7} = $ _____

4. $\frac{7}{12} \times \frac{3}{4} = $ _____

5. $\frac{3}{8} \times \frac{8}{3} = $ _____

6. $\frac{5}{8} \times \frac{9}{10} = $ _____

7. $\frac{6}{7} \times \frac{1}{3} = $ _____

8. $\frac{2}{5} \times \frac{3}{4} = $ _____

9. $\frac{9}{10} \times \frac{1}{3} = $ _____

10. $\frac{7}{10} \times \frac{2}{5} = $ _____

11. $\frac{3}{7} \times \frac{8}{9} = $ _____

12. $\frac{6}{7} \times \frac{14}{17} = $ _____

13. $\frac{7}{28} \times \frac{4}{21} = $ _____

14. $\frac{4}{5} \times \frac{15}{16} = $ _____

15. $\frac{5}{6} \times \frac{18}{25} = $ _____

16. $\frac{3}{16} \times \frac{2}{3} = $ _____

17. $\frac{5}{9} \times \frac{6}{10} = $ _____

18. $\frac{2}{9} \times \frac{3}{10} = $ _____

19. $\frac{5}{8} \times \frac{4}{5} = $ _____

20. $\frac{2}{5} \times \frac{5}{8} = $ _____

21. $\frac{3}{8} \times \frac{4}{9} = $ _____

22. $\frac{8}{15} \times \frac{9}{16} = $ _____

23. $\frac{5}{7} \times \frac{14}{15} = $ _____

24. $\frac{4}{15} \times \frac{5}{8} = $ _____

25. $\frac{3}{5} \times \frac{10}{12} \times \frac{1}{2} = $ _____

26. $\frac{7}{8} \times \frac{4}{7} \times \frac{3}{5} = $ _____

27. $\frac{6}{7} \times \frac{15}{28} \times \frac{14}{27} = $ _____

28. If you took $\frac{3}{5}$ of a cake and then took half of that, how much of the original cake would you have?

Answer: _____

29. If you took $\frac{3}{4}$ of a pizza and then took one-sixth of that, what fractional part of the pizza would you have?

Answer: _____

MULTIPLYING FRACTIONS BY WHOLE NUMBERS

Any whole number can be written as a fraction by inserting a 1 in the denominator.

EXAMPLE   Multiply $3 \times \frac{2}{5}$.   Write 3 as $\frac{3}{1}$. $\frac{3}{1} \times \frac{2}{5} = \frac{6}{5} = 1\frac{1}{5}$

Multiply. Be sure your answer is simplified.

1. $3 \times \frac{1}{2} = $ ___

2. $\frac{2}{3} \times 2 = $ ___

3. $5 \times \frac{3}{4} = $ ___

4. $\frac{1}{2} \times 4 = $ ___

5. $7 \times \frac{3}{10} = $ ___

6. $\frac{3}{4} \times 4 = $ ___

7. $2 \times \frac{3}{5} = $ ___

8. $\frac{1}{3} \times 3 = $ ___

9. $3 \times \frac{1}{4} = $ ___

10. $\frac{7}{10} \times 2 = $ ___

11. $6 \times \frac{4}{5} = $ ___

12. $\frac{3}{8} \times 5 = $ ___

13. $\frac{4}{5} \times 30 = $ ___

14. $50 \times \frac{7}{10} = $ ___

15. $\frac{1}{2} \times 20 = $ ___

16. $\frac{5}{16} \times 400 = $ ___

17. $\frac{7}{12} \times 9 = $ ___

18. $8 \times \frac{5}{12} = $ ___

19. $\frac{7}{12} \times 2 = $ ___

20. $3 \times \frac{3}{5} = $ ___

21. $3 \times \frac{11}{12} = $ ___

22. $\frac{11}{12} \times 2 = $ ___

23. $4 \times \frac{7}{2} = $ ___

24. $\frac{5}{8} \times 6 = $ ___

25. $\frac{5}{3} \times 7 = $ ___

26. $9 \times \frac{2}{3} = $ ___

27. $\frac{7}{20} \times 2 = $ ___

28. $3 \times \frac{7}{15} = $ ___

29. $3 \times \frac{7}{30} = $ ___

30. $\frac{7}{30} \times 2 = $ ___

31. $5 \times \frac{2}{15} = $ ___

32. $\frac{5}{2} \times 4 = $ ___

33. What is two-fifths of $20? _____

34. Find seven-tenths of $80. _____

35. Find three-tenths of $50. _____

36. What is one-eighth of $100? _____

37. What is five-eighths of $40? _____

38. Find three-fifths of $90. _____

39. Find three-fourths of $500. _____

40. Find one-fourth of $5. _____

MULTIPLYING MIXED NUMBERS

Change the mixed number to an improper fraction. Then multiply.

EXAMPLES $2\frac{1}{4} \times \frac{1}{2}$

$\frac{9}{4} \times \frac{1}{2} = \frac{9}{8} = 1\frac{1}{8}$

$\frac{3}{4} \times 5\frac{1}{3}$

$\frac{3}{4} \times \frac{16}{3} = \frac{3}{4} \times \frac{16}{3} = \frac{4}{1} = 4$

Multiply. Then simplify your answer.

1. $3\frac{1}{7} \times \frac{3}{5} =$ ___

2. $\frac{1}{6} \times 5\frac{2}{3} =$ ___

3. $5\frac{3}{4} \times \frac{1}{3} =$ ___

4. $\frac{5}{8} \times 1\frac{3}{4} =$ ___

5. $\frac{2}{5} \times 8\frac{2}{5} =$ ___

6. $11\frac{1}{3} \times \frac{1}{5} =$ ___

7. $\frac{3}{7} \times 9\frac{1}{2} =$ ___

8. $2\frac{3}{7} \times \frac{3}{4} =$ ___

9. $6\frac{1}{3} \times \frac{6}{7} =$ ___

10. $\frac{8}{9} \times 6\frac{3}{5} =$ ___

11. $5\frac{1}{2} \times \frac{6}{7} =$ ___

12. $\frac{5}{6} \times 3\frac{7}{10} =$ ___

13. $\frac{4}{9} \times 7\frac{5}{8} =$ ___

14. $9\frac{2}{7} \times \frac{6}{13} =$ ___

15. $\frac{7}{11} \times 6\frac{7}{8} =$ ___

16. $9\frac{1}{5} \times \frac{8}{23} =$ ___

17. $9\frac{1}{5} \times \frac{5}{8} =$ ___

18. $\frac{7}{9} \times 1\frac{2}{7} =$ ___

19. $3\frac{3}{8} \times \frac{4}{9} =$ ___

20. $\frac{6}{5} \times 4\frac{3}{18} =$ ___

21. $\frac{8}{9} \times 8\frac{1}{4} =$ ___

22. $7\frac{1}{5} \times \frac{5}{9} =$ ___

23. $\frac{12}{13} \times 4\frac{7}{8} =$ ___

24. $12\frac{1}{2} \times \frac{8}{15} =$ ___

25. Anita likes to jog around the track at her school. The track is $\frac{1}{4}$ mile long. Yesterday she jogged around the track $3\frac{1}{2}$ times. How far did she run?

26. The pipe leaked $5\frac{1}{2}$ quarts of water an hour for $7\frac{1}{4}$ hours. How much water leaked in all?

Answer: _____

Answer: _____

MULTIPLYING MIXED NUMBERS

Before you multiply a mixed number by a whole number, convert the mixed number to an improper fraction.

EXAMPLES

$2\frac{1}{2} \times 2$

$\frac{5}{2} \times \frac{2}{1} = \frac{10}{2} = 5$

$3\frac{3}{4} \times 4$

$\frac{11}{3} \times \frac{4}{1} = \frac{44}{3} = 14\frac{2}{3}$

$1\frac{1}{2} \times 2\frac{2}{3}$

$\frac{3}{2} \times \frac{8}{3} = \frac{24}{6} = 4$

Multiply. Then simplify your answer.

1. $1\frac{1}{2} \times 2$

   $\frac{3}{2} \times \frac{2}{1}$

2. $3 \times 2\frac{1}{3}$

3. $3\frac{1}{2} \times 4$

4. $2\frac{2}{3} \times 3$

5. $2 \times 3\frac{1}{5}$

6. $4\frac{1}{2} \times 2$

7. $2\frac{3}{4} \times 2$

8. $4 \times 2\frac{3}{5}$

9. $5\frac{1}{4} \times 3$

10. $3\frac{2}{5} \times 3$

11. $3 \times 4\frac{3}{4}$

12. $2\frac{4}{5} \times 3$

13. $5\frac{1}{3} \times 1\frac{1}{8}$

14. $7\frac{1}{2} \times 2\frac{1}{5}$

15. $6\frac{2}{5} \times 2\frac{3}{8}$

16. If you work for $3\frac{1}{2}$ hours and are paid seven and one-half dollars an hour, how much do you earn?

17. If you buy $4\frac{1}{2}$ yards of cloth for four and a half dollars a yard, how much do you pay?

Answer: _____

Answer: _____

RECIPROCALS

First, write each number as a fraction. Then reverse the positions of numerator and denominator.

EXAMPLE   **Write the reciprocals of $\frac{2}{3}$, 6, and $5\frac{1}{2}$.**

First, write each number as a fraction. $\frac{2}{3} \quad \frac{6}{1} \quad \frac{11}{2}$

Now, reverse the numerator and the denominator. $\frac{3}{2} \quad \frac{1}{6} \quad \frac{2}{11}$

Find the reciprocal of each number.

1. $\frac{1}{4}$ = _____   2. $\frac{2}{5}$ = _____   3. $\frac{3}{8}$ = _____   4. $\frac{9}{5}$ = _____   5. $\frac{7}{9}$ = _____

6. $\frac{6}{11}$ = _____   7. $\frac{7}{13}$ = _____   8. 9 = _____   9. 4 = _____   10. 13 = _____

11. 22 = _____   12. 1 = _____   13. 18 = _____   14. 51 = _____   15. $2\frac{1}{2}$ = _____

16. $1\frac{1}{3}$ = _____   17. $3\frac{2}{3}$ = _____   18. $5\frac{4}{5}$ = _____   19. $8\frac{1}{7}$ = _____   20. $16\frac{2}{5}$ = _____

21. $13\frac{2}{3}$ = _____   22. $\frac{3}{5}$ = _____   23. 7 = _____   24. $2\frac{2}{3}$ = _____   25. $\frac{1}{8}$ = _____

26. 19 = _____   27. $4\frac{1}{5}$ = _____   28. $\frac{9}{2}$ = _____   29. 6 = _____   30. $19\frac{1}{2}$ = _____

31. $\frac{6}{7}$ = _____   32. 15 = _____   33. $13\frac{1}{5}$ = _____   34. $\frac{9}{10}$ = _____   35. 28 = _____

36. $8\frac{1}{3}$ = _____   37. $\frac{1}{6}$ = _____   38. 92 = _____   39. $9\frac{1}{3}$ = _____   40. $\frac{7}{15}$ = _____

41. 29 = _____   42. $11\frac{1}{5}$ = _____   43. $\frac{9}{4}$ = _____   44. $2\frac{3}{7}$ = _____   45. 19 = _____

46. $6\frac{4}{7}$ = _____   47. $\frac{20}{3}$ = _____   48. $8\frac{5}{6}$ = _____   49. 33 = _____   50. $9\frac{1}{6}$ = _____

51. $\frac{6}{19}$ = _____   52. $6\frac{13}{23}$ = _____   53. 100 = _____   54. $8\frac{2}{5}$ = _____   55. $\frac{7}{16}$ = _____

56. It took Janice $9\frac{1}{2}$ days to paint her grandmother's house. About what fraction of the job did she do each day? (Hint: Find the reciprocal of $9\frac{1}{2}$.)

57. Sammy can run once around a track in $2\frac{1}{2}$ minutes. What fraction of a lap can he run in 1 minute?

Answer: _____

Answer: _____

DIVIDING FRACTIONS

First, change the divisor to its reciprocal. Then multiply.

EXAMPLES $\frac{3}{5} \div \frac{7}{8}$ $\frac{5}{9} \div \frac{11}{27}$

$\frac{3}{5} \times \frac{8}{7} = \frac{24}{35}$ $\frac{5}{9} \times \frac{\overset{3}{27}}{11} = \frac{15}{11} = 1\frac{4}{11}$

Divide. Simplify your answer.

1. $\frac{1}{4} \div \frac{3}{5} =$ ___

2. $\frac{3}{5} \div \frac{4}{9} =$ ___

3. $\frac{4}{7} \div \frac{7}{8} =$ ___

4. $\frac{3}{8} \div \frac{1}{3} =$ ___

5. $\frac{2}{5} \div \frac{3}{7} =$ ___

6. $\frac{5}{9} \div \frac{3}{10} =$ ___

7. $\frac{4}{9} \div \frac{3}{4} =$ ___

8. $\frac{7}{12} \div \frac{4}{5} =$ ___

9. $\frac{3}{4} \div \frac{1}{8} =$ ___

10. $\frac{5}{9} \div \frac{2}{3} =$ ___

11. $\frac{3}{4} \div \frac{3}{4} =$ ___

12. $\frac{8}{9} \div \frac{4}{5} =$ ___

13. $\frac{5}{8} \div \frac{1}{2} =$ ___

14. $\frac{9}{13} \div \frac{3}{5} =$ ___

15. $\frac{4}{5} \div \frac{6}{7} =$ ___

16. $\frac{5}{8} \div \frac{1}{4} =$ ___

17. $\frac{9}{10} \div \frac{3}{5} =$ ___

18. $\frac{9}{16} \div \frac{3}{4} =$ ___

19. $\frac{6}{7} \div \frac{9}{14} =$ ___

20. $\frac{9}{11} \div \frac{15}{22} =$ ___

21. $\frac{21}{25} \div \frac{14}{15} =$ ___

22. $\frac{18}{35} \div \frac{3}{10} =$ ___

23. $\frac{22}{25} \div \frac{33}{35} =$ ___

24. $\frac{9}{16} \div \frac{5}{12} =$ ___

25. Mary Ellen has $\frac{3}{10}$ acre for her garden. She needs $\frac{1}{20}$ acre for each vegetable that she plants. How many vegetables can she plant?

Answer: _____

DIVIDING FRACTIONS

First, change the dividend to a fraction. Next, change the divisor to its reciprocal. Then multiply the fractions.

EXAMPLES $1\frac{1}{2} \div \frac{1}{3}$

$$\frac{3}{2} \times \frac{3}{1} = \frac{9}{2} = 4\frac{1}{2}$$

$6 \div \frac{2}{3}$

$$\frac{6}{1} \div \frac{3}{2} = \frac{9}{1} = 9$$

Divide. Simplify your answer.

1. $2\frac{1}{4} \div \frac{1}{3} =$ ___

2. $1\frac{1}{5} \div \frac{2}{3} =$ ___

3. $3\frac{1}{5} \div \frac{4}{7} =$ ___

4. $4\frac{1}{2} \div \frac{3}{4} =$ ___

5. $6 \div \frac{3}{4} =$ ___

6. $8 \div \frac{6}{7} =$ ___

7. $9 \div \frac{7}{8} =$ ___

8. $18 \div \frac{5}{4} =$ ___

9. $5\frac{1}{3} \div \frac{4}{5} =$ ___

10. $16 \div \frac{10}{11} =$ ___

11. $1\frac{7}{12} \div \frac{2}{3} =$ ___

12. $10 \div \frac{15}{16} =$ ___

13. $3\frac{3}{4} \div \frac{7}{12} =$ ___

14. $1\frac{3}{8} \div \frac{3}{4} =$ ___

15. $9 \div \frac{5}{16} =$ ___

16. $2\frac{5}{8} \div \frac{3}{4} =$ ___

17. $4\frac{1}{5} \div \frac{7}{8} =$ ___

18. $8 \div \frac{6}{11} =$ ___

19. $6\frac{2}{3} \div \frac{5}{8} =$ ___

20. $8\frac{2}{3} \div \frac{5}{6} =$ ___

21. $14 \div \frac{16}{17} =$ ___

22. $7\frac{1}{2} \div \frac{5}{6} =$ ___

23. $3\frac{2}{3} \div \frac{2}{3} =$ ___

24. $10 \div \frac{8}{13} =$ ___

DIVIDING MIXED NUMBERS

REMEMBER: First, change both numbers to fractions. Next, change the divisor to its reciprocal. Then multiply the fractions.

EXAMPLES

$1\frac{1}{2} \div 2\frac{1}{3}$

$\downarrow$

$\frac{3}{2} \div \frac{7}{3}$

$\downarrow$

$\frac{3}{2} \times \frac{3}{7} = \frac{9}{14}$

$6 \div 3\frac{3}{7}$

$\downarrow$

$\frac{6}{1} \div \frac{24}{7}$

$\downarrow$

$\frac{6}{1} \times \frac{7}{24} = \frac{7}{4} = 1\frac{3}{4}$

$\frac{3}{8} \div 1\frac{3}{8}$

$\downarrow$

$\frac{3}{8} \div \frac{11}{8}$

$\downarrow$

$\frac{3}{8} \times \frac{8}{11} = \frac{3}{11}$

Divide. Simplify your answer.

1. $2\frac{2}{3} \div 3\frac{1}{3} =$ ___

2. $4\frac{1}{2} \div 5\frac{1}{4} =$ ___

3. $2\frac{1}{2} \div 3\frac{1}{3} =$ ___

4. $5\frac{1}{3} \div 3\frac{2}{3} =$ ___

5. $1 \div 8\frac{3}{4} =$ ___

6. $7 \div 4\frac{9}{10} =$ ___

7. $15 \div 2\frac{1}{7} =$ ___

8. $5 \div 6\frac{2}{3} =$ ___

9. $\frac{1}{2} \div 2\frac{2}{3} =$ ___

10. $\frac{2}{3} \div 2\frac{1}{5} =$ ___

11. $\frac{1}{7} \div \frac{3}{14} =$ ___

12. $\frac{4}{5} \div 6\frac{2}{5} =$ ___

13. $6\frac{3}{4} \div 1\frac{1}{3} =$ ___

14. $12 \div 5\frac{2}{3} =$ ___

15. $3\frac{3}{4} \div 1\frac{1}{4} =$ ___

16. $\frac{7}{9} \div 1\frac{13}{15} =$ ___

17. $2\frac{5}{8} \div 2\frac{1}{4} =$ ___

18. $9 \div 4\frac{1}{8} =$ ___

19. $7\frac{1}{2} \div 3\frac{1}{3} =$ ___

20. $\frac{11}{12} \div 6\frac{1}{9} =$ ___

Solving Equations Using Multiplication and Division

You may multiply or divide both sides of an equation by the same number.

EXAMPLES $5x = 25$ **Divide by 5.** $4y = 10$ **Divide by 4.**

$$x = 5$$

$$y = \frac{10}{4}$$

$$= 2\frac{2}{4} \quad \textbf{Mixed Number}$$

$$= 2\frac{1}{2} \quad \textbf{Simplify}$$

Solve each equation.

1. $2a = 8$

2. $7b = 35$

3. $9c = 27$

4. $6x = 24$

5. $2y = 5$

6. $2z = 9$

7. $10b = 25$

8. $4c = 20$

9. $8a = 20$

10. $3x = 14$

11. $5y = 21$

12. $6x = 16$

13. $4a = 14$

14. $7b = 15$

15. $5c = 12$

16. $9x = 26$

17. $7x = 44$

18. $6z = 47$

SOLVING EQUATIONS USING MULTIPLICATION AND DIVISION

To solve a division equation, you multiply both sides by the same number.

EXAMPLES

$\frac{y}{3} = 5$ **Multiply by 3.**

$3\left(\frac{y}{3}\right) = 3(5)$

$y = 15$

$\frac{a}{2} = 2\frac{1}{2}$ **Multiply by 2.**

$2\left(\frac{a}{2}\right) = 2\left(2\frac{1}{2}\right)$

$a = 5$

Solve each equation.

1. $\frac{x}{2} = 3$

2. $\frac{y}{3} = 4$

3. $\frac{z}{5} = 3$

4. $\frac{a}{3} = 5$

5. $\frac{b}{6} = 3$

6. $\frac{c}{7} = 4$

7. $\frac{x}{2} = 1\frac{1}{2}$

8. $\frac{y}{3} = 1\frac{1}{3}$

9. $2x = 5$

10. $\frac{a}{5} = 10$

11. $3b = 15$

12. $\frac{c}{6} = 2\frac{1}{2}$

13. $\frac{x}{2} = 13$

14. $\frac{y}{4} = 2\frac{1}{2}$

15. $7x = 49$

MULTIPLYING AND DIVIDING FRACTIONS

Richard runs at a rate of 1 lap every $2\frac{3}{4}$ minutes. How long will it take him to run the following distances?

1. 4 laps _____

$(4 \times 2\frac{3}{4})$

2. 5 laps _____

3. 7 laps _____

4. 14 laps _____

5. $2\frac{1}{2}$ laps _____

6. $6\frac{1}{2}$ laps _____

7. $8\frac{1}{4}$ laps _____

8. $9\frac{3}{4}$ laps _____

How many laps can Richard run in the following time periods?

9. 11 minutes _____

$(11 \div 2\frac{3}{4})$

10. $5\frac{1}{2}$ minutes _____

11. $19\frac{1}{2}$ minutes _____

12. 7 minutes _____

Rachel skates at a rate of 1 lap every $2\frac{1}{2}$ minutes. How long will it take her to skate the following distances?

13. 6 laps _____

14. 14 laps _____

15. 9 laps _____

16. 7 laps _____

17. $2\frac{1}{2}$ laps _____

18. $6\frac{1}{2}$ laps _____

19. $3\frac{1}{4}$ laps _____

20. $7\frac{3}{4}$ laps _____

How many laps can Rachel skate in the following time periods?

21. $7\frac{1}{2}$ minutes _____

22. 10 minutes _____

23. $17\frac{1}{2}$ minutes _____

24. 26 minutes _____

FINDING COMMON DENOMINATORS

Write each pair of fractions with a common denominator.

1. $\frac{1}{4}$ and $\frac{2}{3}$ _____

2. $\frac{1}{5}$ and $\frac{2}{3}$ _____

3. $\frac{1}{5}$ and $\frac{3}{8}$ _____

4. $\frac{3}{8}$ and $\frac{3}{4}$ _____

5. $\frac{3}{8}$ and $\frac{13}{16}$ _____

6. $\frac{2}{3}$ and $\frac{1}{6}$ _____

7. $\frac{1}{2}$ and $\frac{2}{3}$ _____

8. $\frac{1}{4}$ and $\frac{5}{6}$ _____

9. $\frac{5}{6}$ and $\frac{5}{12}$ _____

10. $\frac{1}{3}$ and $\frac{2}{5}$ _____

11. $\frac{1}{6}$ and $\frac{2}{9}$ _____

12. $\frac{4}{5}$ and $\frac{1}{2}$ _____

13. $\frac{1}{4}$ and $\frac{3}{5}$ _____

14. $\frac{1}{6}$ and $\frac{3}{8}$ _____

15. $\frac{1}{3}$ and $\frac{7}{10}$ _____

Write these fractions with a common denominator.

16. $\frac{2}{3}$, $\frac{1}{6}$, and $\frac{1}{2}$ _____

17. $\frac{1}{6}$, $\frac{2}{3}$, and $\frac{5}{12}$ _____

18. $\frac{1}{3}$, $\frac{3}{4}$, and $\frac{1}{6}$ _____

19. $\frac{4}{5}$, $\frac{3}{10}$, and $\frac{2}{3}$ _____

88

COMPARING FRACTIONS

First, write the numbers as fractions with a common denominator. Then compare numerators.

EXAMPLE   Which is larger, $\frac{2}{3}$ or $\frac{5}{8}$?
Write the fractions so that they have the same denominator.

$\frac{2}{3} = \frac{16}{24}$ 16 is larger than 15.

$\frac{5}{8} = \frac{15}{24}$ So $\frac{2}{3}$ is larger than $\frac{5}{8}$.

Circle the larger fraction in each pair.

1. $\frac{5}{6}$ $\frac{1}{6}$ 2. $\frac{7}{8}$ $\frac{3}{8}$ 3. $\frac{5}{9}$ $\frac{7}{9}$ 4. $\frac{7}{10}$ $\frac{9}{10}$ 5. $\frac{11}{13}$ $\frac{10}{13}$

6. $\frac{8}{29}$ $\frac{26}{29}$ 7. $\frac{16}{17}$ $\frac{13}{17}$ 8. $\frac{5}{8}$ $\frac{3}{4}$ 9. $\frac{2}{3}$ $\frac{7}{12}$ 10. $\frac{3}{5}$ $\frac{7}{10}$

11. $\frac{1}{3}$ $\frac{4}{15}$ 12. $\frac{5}{9}$ $\frac{2}{3}$ 13. $\frac{3}{4}$ $\frac{11}{12}$ 14. $\frac{2}{7}$ $\frac{5}{21}$ 15. $\frac{1}{3}$ $\frac{3}{5}$

16. $\frac{3}{4}$ $\frac{2}{3}$ 17. $\frac{4}{5}$ $\frac{5}{7}$ 18. $\frac{3}{8}$ $\frac{1}{3}$ 19. $\frac{5}{11}$ $\frac{2}{5}$ 20. $\frac{2}{9}$ $\frac{1}{5}$

21. $\frac{4}{7}$ $\frac{2}{3}$ 22. $\frac{3}{4}$ $\frac{5}{6}$ 23. $\frac{5}{8}$ $\frac{7}{12}$ 24. $\frac{7}{9}$ $\frac{5}{6}$ 25. $\frac{5}{9}$ $\frac{8}{15}$

26. $\frac{3}{10}$ $\frac{1}{4}$ 27. $\frac{5}{14}$ $\frac{8}{21}$ 28. $\frac{3}{8}$ $\frac{5}{14}$ 29. $\frac{5}{9}$ $\frac{3}{5}$ 30. $\frac{6}{11}$ $\frac{5}{9}$

31. $\frac{5}{8}$ $\frac{11}{18}$ 32. $\frac{7}{9}$ $\frac{2}{3}$ 33. $\frac{3}{10}$ $\frac{4}{15}$ 34. $\frac{5}{6}$ $\frac{11}{14}$ 35. $\frac{9}{10}$ $\frac{3}{5}$

ADDING FRACTIONS

REMEMBER: To add fractions with like denominators, add the numerators. The denominator stays the same.

EXAMPLES

$$\frac{1}{3} + \frac{1}{3} = \frac{2}{3}$$

$$\frac{1}{5} + \frac{2}{5} = \frac{3}{5}$$

$$\frac{2}{6} + \frac{1}{6} = \frac{3}{6} = \frac{1}{2}$$

$$\frac{1}{4} + \frac{3}{4} = \frac{4}{4} = 1$$

Add. Reduce if possible.

1. $\frac{1}{4} + \frac{2}{4}$

2. $\frac{2}{5} + \frac{2}{5}$

3. $\frac{1}{6} + \frac{3}{6}$

4. $\frac{3}{10} + \frac{4}{10}$

5. $\frac{2}{12} + \frac{3}{12}$

6. $\frac{1}{10} + \frac{3}{10}$

7. $\frac{3}{12} + \frac{5}{12}$

8. $\frac{2}{10} + \frac{5}{10}$

9. $\frac{1}{4} + \frac{3}{4}$

10. $\frac{2}{6} + \frac{1}{6}$

11. $\frac{2}{20} + \frac{3}{20}$

12. $\frac{1}{8} + \frac{3}{8}$

13. $\frac{2}{8} + \frac{4}{8}$

14. $\frac{5}{10} + \frac{1}{10}$

15. $\frac{1}{12} + \frac{11}{12}$

16. $\frac{4}{10} + \frac{5}{10}$

17. $\frac{3}{12} + \frac{3}{12}$

18. $\frac{1}{20} + \frac{10}{20}$

19. $\frac{2}{12} + \frac{6}{12}$

20. $\frac{3}{20} + \frac{17}{20}$

21. $\frac{3}{7} + \frac{2}{7}$

22. $\frac{5}{12} + \frac{7}{12}$

23. $\frac{7}{25} + \frac{15}{25}$

24. $\frac{11}{50} + \frac{27}{50}$

25. $\frac{4}{15} + \frac{9}{15}$

ADDING FRACTIONS

REMEMBER: When adding fractions with different denominators, first find a common denominator. The least common denominator (LCD) is the smallest number that can be evenly divided by each denominator.

EXAMPLES

$$\frac{1}{2} \longrightarrow \frac{3}{6}$$
$$+\frac{1}{3} \longrightarrow +\frac{2}{6}$$
$$\frac{5}{6}$$

$$\frac{1}{4} \longrightarrow \frac{3}{12}$$
$$+\frac{2}{3} \longrightarrow +\frac{8}{12}$$
$$\frac{11}{12}$$

Add.

1. $\frac{1}{3} \longrightarrow$
   $+\frac{1}{4} \longrightarrow$

2. $\frac{1}{3} \longrightarrow$
   $+\frac{2}{4} \longrightarrow$

3. $\frac{1}{2}$
   $+\frac{1}{3}$

4. $\frac{1}{5}$
   $+\frac{1}{3}$

5. $\frac{1}{6}$
   $+\frac{4}{15}$

6. $\frac{2}{3}$
   $+\frac{1}{4}$

7. $\frac{1}{5}$
   $+\frac{1}{4}$

8. $\frac{2}{5}$
   $+\frac{1}{3}$

9. $\frac{3}{10}$
   $+\frac{1}{5}$

10. $\frac{3}{10}$
    $+\frac{1}{6}$

11. $\frac{1}{4}$
    $+\frac{7}{10}$

12. $\frac{1}{6}$
    $+\frac{1}{4}$

13. $\frac{2}{6}$
    $+\frac{1}{2}$

14. $\frac{2}{3}$
    $+\frac{1}{5}$

15. $\frac{5}{8}$
    $+\frac{1}{12}$

16. $\frac{9}{10}$
    $+\frac{1}{12}$

17. $\frac{5}{6}$
    $+\frac{1}{10}$

18. $\frac{3}{5}$
    $+\frac{2}{6}$

19. $\frac{3}{10}$
    $+\frac{4}{15}$

20. $\frac{1}{8}$
    $+\frac{3}{20}$

SUBTRACTING FRACTIONS

To subtract one fraction from another, you must have like denominators. When denominators are like, you subtract one numerator from another.

EXAMPLES

$$\frac{\frac{2}{5}}{-\frac{1}{5}} \over \frac{1}{5}$$

$$\frac{3}{4} \longrightarrow \frac{9}{12}$$
$$-\frac{2}{3} \longrightarrow -\frac{8}{12} \over \frac{1}{12}$$

$$\frac{4}{5} \longrightarrow \frac{12}{15}$$
$$-\frac{2}{3} \longrightarrow -\frac{10}{15} \over \frac{2}{15}$$

Subtract.

1. $\frac{2}{3}$
  $-\frac{1}{3}$

2. $\frac{8}{15}$
  $-\frac{4}{15}$

3. $\frac{7}{12}$
  $-\frac{5}{12}$

4. $\frac{6}{10}$
  $-\frac{1}{10}$

5. $\frac{4}{4}$
  $-\frac{1}{4}$

6. $\frac{5}{6}$
  $-\frac{3}{6}$

7. $\frac{19}{20}$
  $-\frac{1}{20}$

8. $\frac{3}{4} \longrightarrow$
  $-\frac{1}{2} \longrightarrow$

9. $\frac{3}{5} \longrightarrow$
  $-\frac{1}{4} \longrightarrow$

10. $\frac{2}{3} \longrightarrow$
  $-\frac{1}{5} \longrightarrow$

11. $\frac{5}{6} \longrightarrow$
  $-\frac{2}{3} \longrightarrow$

12. $\frac{1}{2} \longrightarrow$
  $-\frac{1}{5} \longrightarrow$

13. $\frac{3}{8} \longrightarrow$
  $-\frac{1}{4} \longrightarrow$

14. $\frac{5}{6}$
  $-\frac{3}{4}$

15. $\frac{9}{10}$
  $-\frac{3}{4}$

16. $\frac{11}{20}$
  $-\frac{1}{2}$

17. $\frac{7}{8}$
  $-\frac{1}{2}$

18. $\frac{3}{4}$
  $-\frac{2}{5}$

19. $\frac{9}{10}$
  $-\frac{3}{5}$

ADDING MIXED NUMBERS

REMEMBER: If the fractions need to be renamed, rewrite the whole numbers also.

EXAMPLES

$2\frac{1}{5}$
$+3\frac{3}{5}$
$\overline{5\frac{4}{5}}$

$6\frac{1}{2} \rightarrow 6\frac{3}{6}$
$+2\frac{1}{3} \rightarrow 2\frac{2}{6}$
$\overline{8\frac{5}{6}}$

$5\frac{1}{2} + 2\frac{1}{4}$

$5\frac{1}{2} \rightarrow 5\frac{2}{4}$
$+2\frac{1}{4} \rightarrow 2\frac{1}{4}$
$\overline{7\frac{3}{4}}$

Add.

1. $2\frac{3}{8}$
$+5\frac{4}{8}$

2. $8\frac{1}{3}$
$+6\frac{1}{3}$

3. $2\frac{1}{8}$
$+4\frac{3}{8}$

4. $9\frac{3}{16}$
$+4\frac{7}{16}$

5. $7\frac{1}{6}$
$+3\frac{3}{6}$

6. $8\frac{2}{3}$
$+7\frac{1}{6}$

7. $3\frac{1}{4}$
$+6\frac{1}{2}$

8. $5\frac{3}{8}$
$+6\frac{1}{4}$

9. $4\frac{2}{5}$
$+3\frac{1}{2}$

10. $6\frac{1}{4}$
$+3\frac{2}{3}$

11. $5\frac{1}{3}$
$+6\frac{3}{8}$

12. $9\frac{3}{5}$
$+9\frac{1}{4}$

13. $1\frac{1}{2}$
$+2\frac{1}{3}$

14. $7\frac{1}{3}$
$+8\frac{2}{5}$

15. $6\frac{1}{6}$
$+3\frac{3}{8}$

16. $5\frac{1}{8} + 6\frac{5}{6} = $ _____

17. $6\frac{1}{8} + 3\frac{3}{4} = $ _____

18. $4\frac{1}{3} + 5\frac{1}{12} = $ _____

19. $1\frac{1}{7} + 3\frac{2}{3} = $ _____

20. $6\frac{1}{2} + 1\frac{2}{7} = $ _____

21. $5\frac{1}{3} + 4\frac{1}{6} = $ _____

ADDING MIXED NUMBERS

Add.

1. $2\frac{3}{7}$
   $+\ \frac{2}{7}$

2. $9\frac{1}{3}$
   $+\ 6$

3. $\frac{5}{6}$
   $+\ 9$

4. $5\frac{3}{8}$
   $+\ 1\frac{3}{8}$

5. $\frac{1}{6}$
   $+\ 3\frac{1}{4}$

6. 2
   $+\ 7\frac{5}{8}$

7. 16
   $+\ \frac{3}{5}$

8. $5\frac{1}{2}$
   $+\ 3\frac{3}{8}$

9. $\frac{3}{4}$
   $+\ 9\frac{3}{16}$

10. 6
    $+\ 9\frac{5}{12}$

11. $6\frac{1}{2}$
    3
    $+\ 8\frac{1}{4}$

12. $2\frac{1}{8}$
    $\frac{1}{4}$
    $+\ 6\frac{1}{2}$

13. 16
    $2\frac{3}{10}$
    $+\ 4\frac{1}{5}$

14. $8\frac{1}{8}$
    $6\frac{1}{6}$
    $+\ \frac{1}{4}$

15. $9\frac{1}{3}$
    $\frac{1}{5}$
    $+\ 6$

16. $6 + 4\frac{1}{7} =$ _____

17. $5\frac{1}{3} + \frac{1}{9} =$ _____

18. $7\frac{1}{2} + 9\frac{1}{6} =$ _____

19. $\frac{5}{16} + 3\frac{1}{2} =$ _____

20. $8 + \frac{7}{12} =$ _____

21. $5\frac{3}{4} + 2\frac{1}{8} =$ _____

ADDING MIXED NUMBERS

First, rewrite the fractions with a common denominator. Next, add the whole numbers. Then add the fractions. Then convert the sum to a proper mixed number or a whole number.

EXAMPLES

$$5\frac{3}{4} \longrightarrow 5\frac{9}{12}$$
$$+\ 1\frac{2}{3} \longrightarrow 1\frac{8}{12}$$
$$6\frac{17}{12} = 7\frac{5}{12}$$

$$2\frac{1}{3} \longrightarrow 2\frac{2}{6}$$
$$+\ 3\frac{4}{6} \longrightarrow 3\frac{4}{6}$$
$$5\frac{6}{6} = 6$$

Add.

1. $6\frac{1}{3}$
   $+\ 3\frac{3}{4}$

2. $6\frac{5}{8}$
   $+\ 5\frac{2}{3}$

3. $7\frac{7}{12}$
   $+\ 2\frac{5}{6}$

4. $3\frac{2}{5}$
   $+\ 9\frac{6}{10}$

5. $8\frac{3}{4}$
   $+\ 7\frac{4}{5}$

6. $6\frac{7}{10}$
   $+\ 5\frac{2}{3}$

7. $5\frac{1}{6}$
   $+\ 4\frac{7}{8}$

8. $5\frac{7}{8}$
   $+\ 2\frac{3}{5}$

9. $1\frac{3}{8}$
   $+\ 6\frac{15}{24}$

10. $2\frac{5}{12}$
    $+\ 4\frac{5}{6}$

11. $9\frac{2}{3}$
    $+\ 5\frac{7}{10}$

12. $3\frac{3}{4}$
    $+\ 1\frac{2}{5}$

13. $6\frac{1}{2} + 9\frac{5}{7} =$ _____

14. $6\frac{5}{6} + 2\frac{5}{8} =$ _____

15. $9\frac{7}{10} + \frac{5}{8} =$ _____

OPERATIONS WITH MIXED NUMBERS

Harvey works as a mechanic at Mac's Garage. He keeps a weekly record of the number of hours he works each day. For each week, find the total number of hours worked. Then multiply his total hours for that week by $24 per hour to find his earnings for that week.

EXAMPLE

Day	Hours Worked
Monday	$7\frac{1}{2}$
Tuesday	$7\frac{1}{4}$
Wednesday	$5\frac{1}{2}$
Thursday	$8\frac{1}{2}$
Friday	$7\frac{3}{4}$
Total	$36\frac{1}{2}$

Now multiply.

$\frac{24}{1} \times \frac{73}{2} = \$876$

1.

Day	Hours Worked
Monday	5
Tuesday	$6\frac{1}{2}$
Wednesday	6
Thursday	$8\frac{1}{2}$
Friday	$4\frac{1}{2}$
Total	_____
Earnings	_____

2.

Day	Hours Worked
Monday	$3\frac{1}{2}$
Tuesday	$5\frac{1}{2}$
Wednesday	$9\frac{1}{4}$
Thursday	$4\frac{1}{4}$
Friday	$3\frac{1}{4}$
Total	_____
Earnings	_____

3.

Day	Hours Worked
Monday	$4\frac{1}{4}$
Tuesday	$4\frac{1}{4}$
Wednesday	$4\frac{2}{3}$
Thursday	$8\frac{1}{3}$
Friday	$3\frac{1}{3}$
Saturday	$6\frac{3}{4}$
Total	_____
Earnings	_____

4.

Day	Hours Worked
Monday	$6\frac{3}{4}$
Tuesday	$6\frac{1}{2}$
Wednesday	$5\frac{1}{2}$
Thursday	$6\frac{3}{4}$
Friday	$4\frac{1}{4}$
Saturday	7
Total	_____
Earnings	_____

5.

Day	Hours Worked
Monday	$7\frac{1}{2}$
Tuesday	5
Wednesday	$6\frac{1}{3}$
Thursday	$4\frac{3}{4}$
Friday	$3\frac{5}{6}$
Saturday	$5\frac{1}{3}$
Total	_____
Earnings	_____

SUBTRACTING MIXED NUMBERS

REMEMBER: When subtracting mixed numbers, subtract the fraction from the fraction and the whole number from the whole number. Be sure that the fractions have a common denominator. Be careful to line up whole number under whole number and fraction under fraction.

EXAMPLES

$3\frac{1}{2}$ ⟶ $3\frac{3}{6}$ $5\frac{3}{4}$ ⟶ $5\frac{9}{12}$

$-\ 1\frac{1}{3}$ ⟶ $-\ 1\frac{2}{6}$ $-\ 2\frac{2}{3}$ ⟶ $-\ 2\frac{8}{12}$

$2\frac{1}{6}$ $3\frac{1}{12}$

Subtract.

1. $4\frac{4}{19}$ ⟶
$-\ 2\frac{3}{10}$ ⟶

2. $3\frac{1}{2}$ ⟶
$-\ 2\frac{1}{4}$ ⟶

3. $5\frac{3}{4}$ ⟶
$-\ 3\frac{1}{3}$ ⟶

4. $7\frac{1}{2}$ ⟶
$-\ 2\frac{1}{5}$ ⟶

5. $7\frac{3}{8}$
$-\ 2\frac{1}{4}$

6. $6\frac{3}{5}$
$-\ 2\frac{1}{4}$

7. $8\frac{5}{6}$
$-\ 5\frac{1}{3}$

8. $6\frac{2}{3}$
$-\ 2\frac{1}{5}$

9. $5\frac{1}{2}$
$-\ 4\frac{1}{6}$

10. $9\frac{4}{5}$
$-\ 5\frac{1}{3}$

11. $10\frac{3}{4}$
$-\ 2\frac{3}{10}$

12. $9\frac{2}{3}$
$-\ 6\frac{1}{4}$

13. $2\frac{7}{8}$
$-\ 1\frac{1}{2}$

14. $3\frac{3}{5}$
$-\ 1\frac{6}{10}$

15. $2\frac{2}{3}$
$-\ 2\frac{1}{2}$

16. $7\frac{5}{6}$
$-\ 2\frac{1}{4}$

17. $9\frac{1}{5}$
$-\ 3\frac{1}{10}$

18. $8\frac{5}{8}$
$-\ 1\frac{1}{2}$

19. $6\frac{5}{6}$
$-\ 2\frac{1}{8}$

20. $5\frac{2}{3}$
$-\ 5\frac{1}{5}$

SUBTRACTING MIXED NUMBERS

REMEMBER: You can borrow 1 from a whole number and convert it into a fraction with equal numerator and denominator.

EXAMPLES

$$2 \longrightarrow 1\frac{2}{2}$$
$$-\ \frac{1}{2} \longrightarrow -\ \frac{1}{2}$$
$$\overline{1\frac{1}{2}}$$

$$9 \longrightarrow 8\frac{5}{5}$$
$$-\ 3\frac{2}{5} \longrightarrow -\ 3\frac{2}{5}$$
$$\overline{5\frac{3}{5}}$$

Subtract.

1. 5
    $-\ \frac{2}{3}$

2. 6
    $-\ \frac{3}{5}$

3. 3
    $-\ \frac{1}{2}$

4. 2
    $-\ \frac{1}{3}$

5. 5
    $-\ \frac{4}{5}$

6. 5
    $-\ 3\frac{6}{10}$

7. 10
    $-\ 3\frac{2}{3}$

8. 12
    $-\ 5\frac{7}{10}$

9. 11
    $-\ 2\frac{3}{20}$

10. 5
    $-\ 4\frac{3}{8}$

11. 6
    $-\ \frac{5}{8}$

12. 6
    $-\ 2\frac{3}{4}$

13. 8
    $-\ \frac{3}{5}$

14. 10
    $-\ 9\frac{5}{6}$

15. 9
    $-\ \frac{4}{6}$

16. 8
    $-\ 3\frac{1}{3}$

17. 5
    $-\ \frac{1}{12}$

18. 9
    $-\ 2\frac{1}{6}$

19. 1
    $-\ \frac{4}{15}$

20. 8
    $-\ 6\frac{1}{2}$

21. The Walters bought 5 pounds of sugar. They used $1\frac{1}{4}$ pounds to do their holiday baking. How much sugar was left?

Answer: _____

98

SUBTRACTING MIXED NUMBERS

EXAMPLES

$3\frac{1}{2} \longrightarrow 3\frac{3}{6} \longrightarrow 2\frac{9}{6}$

$-1\frac{2}{3} \longrightarrow -1\frac{4}{6} \longrightarrow -1\frac{4}{6}$

$\phantom{-1\frac{4}{6} \longrightarrow} 1\frac{5}{6}$

$5\frac{2}{5} \longrightarrow 5\frac{8}{20} \longrightarrow 4\frac{28}{20}$

$-2\frac{3}{4} \longrightarrow -2\frac{15}{20} \longrightarrow -2\frac{15}{20}$

$\phantom{-2\frac{15}{20} \longrightarrow} 2\frac{13}{20}$

Subtract.

1. $\quad 5\frac{1}{2}$
   $-2\frac{4}{5}$

2. $\quad 4\frac{3}{10}$
   $-2\frac{1}{2}$

3. $\quad 2\frac{1}{2}$
   $-1\frac{3}{4}$

4. $\quad 6\frac{1}{3}$
   $-2\frac{3}{5}$

5. $\quad 7\frac{1}{4}$
   $-3\frac{4}{5}$

6. $\quad 10\frac{2}{6}$
   $-\ 4\frac{1}{2}$

7. $\quad 5\frac{1}{10}$
   $-3\frac{3}{4}$

8. $\quad 6\frac{5}{8}$
   $-2\frac{3}{4}$

9. $\quad 5\frac{2}{3}$
   $-4\frac{4}{5}$

10. $\quad 6\frac{1}{6}$
    $-2\frac{3}{4}$

11. $\quad 7\frac{1}{2}$
    $-6\frac{5}{6}$

12. $\quad 5\ \frac{2}{5}$
    $-1\ \frac{3}{4}$

13. $\quad 8\frac{1}{2}$
    $-3\frac{7}{8}$

14. $\quad 6\frac{1}{6}$
    $-3\frac{5}{8}$

15. $\quad 10\ \frac{1}{3}$
    $-\ 9\frac{7}{10}$

OPERATIONS WITH FRACTIONS

Read each exercise carefully. Decide which operation to use.

EXAMPLES

If you started with $9 and spent two-thirds of it, how much money did you spend?

$$\overset{3}{\cancel{9}} \times \frac{2}{\underset{1}{\cancel{3}}} = 6 \longrightarrow \$6$$

If you spent one-third of your money for lunch and one-half for dinner, what part did you spend in all?

Lunch	$\frac{1}{3}$	$\frac{2}{6}$
Dinner	$+\frac{1}{2}$	$+\frac{3}{6}$
Total		$\frac{5}{6}$

---

1. If you spent one-half of your money for a book and one-fifth for a movie, what part did you spend in all?

Answer: _____

2. If, in exercise 1, you started with $20, how much did you spend?

Answer: _____

---

3. A tennis racket costs two-thirds the amount of a pair of skis. The skis cost $60. Find the cost of the tennis racket.

Answer: _____

4. If you spent one-fourth of your money for a shirt and two-thirds of your money for a sweater, what fraction of your money did you spend?

Answer: _____

---

5. If, in exercise 4, you had $30, how much would it cost to buy both the shirt and the sweater?

Answer: _____

6. If you started with $100 and spent one-half for clothes and one-fourth for shoes, how much money would you have left?

Answer: _____

OPERATIONS WITH FRACTIONS

Read each problem carefully. Decide which operation you should use. There are 24 hours in a day, 12 months in a year, and 60 minutes in an hour.

EXAMPLES

If you spend $\frac{1}{4}$ day in school, how many hours is it?

$$\overset{6}{\cancel{24}} \times \frac{1}{\underset{1}{\cancel{4}}} = 6 \text{ hours}$$

If you were $\frac{3}{4}$ hour late, how many minutes were you late?

$$60 \times \frac{3}{4} = 45 \text{ minutes}$$

---

1. If you waited $\frac{2}{3}$ hour for someone, how long did you wait?

   Answer: _____

2. If you spent $\frac{1}{12}$ year at camp and $\frac{1}{6}$ traveling, for what fraction of a year were you away?

   Answer: _____

---

3. In Exercise 2, for how many months were you away?

   Answer: _____

4. If you ran for $\frac{2}{3}$ hour and walked for $1\frac{1}{2}$ hours, how long did you travel?

   Answer: _____

---

5. If you slept for $\frac{1}{3}$ day and spent $\frac{1}{12}$ day eating, what part did you spend in eating and sleeping?

   Answer: _____

6. In Exercise 5, how many hours did you spend eating and sleeping?

   Answer: _____

---

7. If you spent $3\frac{1}{3}$ hours riding a bike and $2\frac{1}{3}$ hours playing ball, how many hours did you spend doing both?

   Answer: _____

8. How many minutes are there in $\frac{5}{12}$ hour?

   Answer: _____

OPERATIONS WITH FRACTIONS

REMEMBER: Read each problem carefully. Decide which information to use. Choose the right operation.

A family took a trip in which they traveled 360 miles in 6 days and spent $1080. Use this information to solve the following exercises.

EXAMPLE
If they traveled one-third of the distance on the first day, how far did they go?

$$\frac{1}{\overset{}{\underset{1}{3}}} \times \frac{\overset{120}{\cancel{360}}}{1} = 120 \text{ miles}$$

1. If they used one-fourth of their money for food, how much was that?

    Answer: _____

2. If they went five-twelfths of the trip on the last day, how far was that?

    Answer: _____

3. What would be one-fourth of the total time for the trip in hours?

    Answer: _____

4. If they spent one-twelfth of their money on entertainment, how much was that?

    Answer: _____

5. After they spent nine-tenths of their money, how much money would be left?

    Answer: _____

6. If they spent one-fifth of their money on one day and one-sixth the next day, how much did they spend in those two days?

    Answer: _____

7. What would be one-ninth of the total amount of money spent?

    Answer: _____

102

OPERATIONS WITH FRACTIONS

REMEMBER: Read each problem carefully. Think about what you know and what you are to find out. Do the first step. Then do the second step.

Attendance at a high school basketball game was 360. Use this information and the additional information given to solve the problems.

EXAMPLE
If one-fourth of the people each bought one hot dog for $0.75, how much did they spend on hot dogs?

$$\frac{1}{\cancel{4}} \times \frac{\cancel{360}^{90}}{1} = 90 \text{ people}$$

$$\begin{array}{r} \$0.75 \\ \times \quad 90 \\ \hline \$67.50 \quad \textbf{Answer} \end{array}$$

1. If one-third of those attending were female, how many were male?

Answer: _____

---

2. If one-half of the people paid $1.50 admission, what is the total amount they paid?

Answer: _____

3. If one-twelfth of the people came late, how many were on time?

Answer: _____

---

4. If three-eighths of the people bought drinks at $0.80 each, what is the total amount they paid?

Answer: _____

5. If two-ninths of the people bought banners for $2.50 each, what is the total amount they spent?

Answer: _____

---

6. If three-twelfths of the people brought cars and paid $0.50 each to park, what is the total amount they paid to park?

Answer: _____

7. If one-sixth of the people left early, how many stayed until the end?

Answer: _____

OPERATIONS WITH FRACTIONS

Think about what you know and what you must find out. Do the first step. Then do the second step.

Attendance at a college football game was 7200. Use this and the additional information given to solve the problems.

EXAMPLE
If one-third of the people paid thirty-five cents each to come by bus, how much did they pay in all?

$$\frac{1}{\overset{1}{\cancel{3}}} \times \frac{\overset{2400}{\cancel{7200}}}{1} = \frac{2400}{1}$$

$$\begin{array}{r} 2400 \\ \times \quad .35 \\ \hline \$840.00 \quad \textbf{Answer} \end{array}$$

1. If two-fifths of the people paid $5 for their tickets, how much did they pay in all?

Answer: _____

---

2. If one-sixth of the people bought buttons for $0.60 each, how much did they spend in all?

Answer: _____

3. If five-eighths of the people bought programs for $1.25 each, how much did they spend in all?

Answer: _____

---

4. If one-ninth of the people came late, how many people were on time?

Answer: _____

5. If one-tenth of the people wore blue hats, how many did not wear blue hats?

Answer: _____

---

6. If seven-twelfths of the people rooted for the home team, how many did not?

Answer: _____

7. If one-sixtieth of the people were band members and half of the band were female, how many band members were female?

Answer: _____

104

Sometimes, problems with fractions or mixed numbers are hard to solve. Try to think about a similar problem with whole numbers first.

EXAMPLES **Janice can run once around a track in $2\frac{1}{4}$ minutes. How long will it take her to run around the track four times?**

First: If you use 2 minutes instead of $2\frac{1}{4}$ it takes about $4 \times 2 = 8$ or $2 + 2 + 2 + 2 = 8$ minutes.

Then solve: Use $2\frac{1}{4}$ to solve the problem the same way.

It takes $4 \times 2\frac{1}{4} = 9$ or $2\frac{1}{4} + 2\frac{1}{4} + 2\frac{1}{4} + 2\frac{1}{4} = 9$ minutes.

Derek can walk a mile in about $14\frac{1}{2}$ minutes. He walks $3\frac{1}{10}$ miles to school each morning. Does that take more than an hour?

First: If you use 14 minutes instead of $14\frac{1}{2}$, and 3 miles instead of $3\frac{1}{10}$, it takes about $3 \times 14 = 42$ minutes.

Then solve: 42 minutes is much less than an hour, so you don't need to compute with mixed numbers to answer the question. It takes less than one hour.

---

1. Marco's schoolbooks weigh $3\frac{3}{4}$ lb, $2\frac{1}{2}$ lb, $3\frac{1}{2}$ lb, and $3\frac{1}{4}$ lb. If he carries all of them in his backpack at once, is he carrying more than 10 pounds?

Answer: _____

2. Roberto's son was 3 feet tall in January 1998. He grew $3\frac{1}{4}$ inches in 1998. If he continues to grow $3\frac{1}{4}$ inches per year, how tall will he be in January 2002 (3 more years)?

Answer: _____

---

3. Ms. Shelley lives $17\frac{1}{2}$ miles from her job. If she drives round trip to work 5 days a week, how many miles does she drive weekly?

Answer: _____

4. One batch of fruit drink makes $6\frac{3}{4}$ cups. If Ishmael needs to make at least 16 cups (1 gallon) of fruit drink, how many batches does he need to make?

Answer: _____

5. Three years ago, the fir tree outside Donna's bedroom window was only $6\frac{1}{2}$ feet high. Now, the tree is 12 feet high. How much did it grow in the last three years?

Answer: _____

6. A recipe for fruit juice needs $2\frac{1}{2}$ cups of orange juice, $1\frac{1}{4}$ cups of grape juice, 2 cups of berries, and $\frac{3}{4}$ cup of pineapple juice. Will it all fit into a pitcher that holds 8 cups ($\frac{1}{2}$ gallon)?

Answer: _____

7. Mrs. Jamison can walk 1 mile in $12\frac{3}{4}$ minutes. If she continues at the same pace, can she walk 5 miles in less than 1 hour?

Answer: _____

8. Shareefa bought some stock for $10\frac{1}{2}$ dollars per share. She sold the stock for $18\frac{1}{4}$ dollars per share. What was her profit in dollars per share?

Answer: _____

9. Vera swims 5 days a week. This week she swam $2\frac{1}{2}$ miles, $2\frac{1}{4}$ miles, 3 miles, 2 miles, and $2\frac{3}{4}$ miles. How far did she swim in all?

Answer: _____

10. Rodney owns $13\frac{1}{2}$ acres of land. He wants to split it into two equal parts and sell one of the two lots. How big will each lot be?

Answer: _____

11. Michael was $21\frac{1}{2}$ inches long when he was born. At the age of 20 he was $69\frac{3}{4}$ inches tall. How much did he grow in 20 years?

Answer: _____

12. Jaime measured the rainfall for one year. Winter had $13\frac{7}{10}$ inches, spring had $11\frac{3}{10}$ inches, summer had 5 inches, and fall had $9\frac{9}{10}$ inches. Which two seasons had the most rainfall?

Answer: _____

106

CHAPTER 3

POSTTEST

1. Write the fraction that matches the shaded part.

 _____

2. Shade the figure to match the given fraction.

$\frac{7}{10}$

3. Draw a figure and shade it to match the fraction.

$\frac{3}{5}$

4. Reduce to lowest terms.

$\frac{9}{12} =$ _____

5. Convert to an equivalent fraction with the denominator 16.

$\frac{3}{4} = \frac{}{16}$

6. Convert to a mixed number.

$\frac{90}{12} =$ _____

7. Convert to an improper fraction.

$4\frac{7}{8} =$ _____

8.
$$\begin{array}{r} \frac{3}{6} \\ + \frac{1}{6} \\ \hline \end{array}$$

9.
$$\begin{array}{r} \frac{3}{8} \\ + \frac{5}{6} \\ \hline \end{array}$$

10.
$$\begin{array}{r} 3\frac{4}{5} \\ + 2\frac{3}{5} \\ \hline \end{array}$$

11. $\frac{9}{16} - \frac{3}{16} =$ _____

12. $\frac{5}{8} - \frac{1}{3} =$ _____

13.
$$\begin{array}{r} 9\frac{3}{4} \\ - 1\frac{3}{8} \\ \hline \end{array}$$

14.
$$\begin{array}{r} 4\frac{1}{3} \\ - 2\frac{2}{3} \\ \hline \end{array}$$

15. $\frac{3}{8} \times \frac{4}{9} =$ _____

16. $12 \times \frac{3}{4} =$ _____

17. $5\frac{4}{5} \times \frac{5}{6} =$ _____

18. $\frac{5}{12} \div \frac{3}{4} =$ _____

19. $3\frac{1}{4} \div \frac{3}{8} =$ _____

20. $\frac{7}{12} \div 4 =$ _____

21. $\frac{3}{4} \div 5\frac{1}{6} =$ _____

22. If you started with $16 and spent one-fourth of it, how much did you spend?

Answer: _____

23. How many minutes is $\frac{5}{6}$ hour?

Answer: _____

24. If you spent one-half of your money on Monday, and one-third on Tuesday, what part did you spend in all?

Answer: _____

25. If a woman had $750 and spent one-third of it, how much did she have left?

Answer: _____

26. If 100 people were in a theater and 20 left early, what fraction left early?

Answer: _____

27. If 80 people were at a game and one-fourth bought buttons for $0.75 each, how much money was spent on buttons?

Answer: _____

28. If you had $120 and lost one-third of it, how much money would you have left?

Answer: _____

29. How many hours are in $\frac{1}{6}$ day?

Answer: _____

30. Convert $\frac{5}{8}$ to a decimal.

Answer: _____

31. Convert .013 to a fraction.

Answer: _____

32. Solve for y. $6y = 14$

Answer: _____

33. Solve for c. $\frac{c}{2} = 2\frac{1}{2}$

Answer: _____

PRETEST

Convert the following percents to decimals.

1. 60% = _____

2. 350% = _____

3. $6\frac{1}{2}\%$ = _____

4. 2.53% = _____

5. 1% = _____

6. 15% = _____

Convert the following percents to fractions. Then reduce.

7. 25% = _____

8. 210% = _____

9. $5\frac{1}{2}\%$ = _____

Convert the following to percents.

10. .68 = _____

11. 29 = _____

12. .305 = _____

13. 3.5 = _____

14. .06 = _____

15. .4 = _____

Convert the following to percents.

16. $\frac{1}{4}$ = _____

17. $\frac{1}{6}$ = _____

18. $\frac{3}{8}$ = _____

19. 5 = _____

20. Find 25% of 120. _____

21. Find 6.5% of $3.20. _____

22. 5 is what percent of 10? _____

23. 2.5 is what percent of 12.5? _____

24. If you spent $4 out of $20, what percent did you spend?

Answer: _____

25. 15 is 5% of what number?

Answer: _____

26. If a $35 item is sold at a 10% discount, what is the new selling price?

Answer: _____

27. Find the total price of a $125.50 item with 3% sales tax.

Answer: _____

28. What is the price of a chair that usually sells for $150, but is marked "15% off?"

Answer: _____

29. A house worth $45,000 3 years ago is now worth 32% more. What is it worth now?

Answer: _____

30. If 35 miles is 25% of a total distance, what is the distance?

Answer: _____

31. Find the interest on $1200 at 5% for 1 year.

Answer: _____

32. Find the principal if the interest at 6% is $18 for 1 year.

Answer: _____

33. If a car costs $6500 and you must make a 20% down payment, what is the down payment?

Answer: _____

110

WRITING PERCENTS AS DECIMALS

REMEMBER: You convert a percent to a decimal by dividing by 100. (Hint: To divide by 100, move the decimal point two places to the left.)

EXAMPLES **36% = .36**

$\frac{1}{2}$**% = .5% = .005**

100% = 1.00, or 1

8% = .08

3.6% = .036

.19% = .0019

Convert to decimals.

1. 60% = _____
2. 18% = _____
3. 5% = _____
4. 92% = _____

5. 75% = _____
6. 7% = _____
7. 15% = _____
8. 300% = _____

9. 2% = _____
10. 5000% = _____
11. 68% = _____
12. 115% = _____

13. 840% = _____
14. 41% = _____
15. 1000% = _____
16. 3% = _____

17. 201% = _____
18. 223% = _____
19. 1% = _____
20. 51% = _____

21. 5.4% = _____
22. .93% = _____
23. 12.3% = _____
24. .0008% = _____

25. 2.8% = _____
26. 8.6% = _____
27. 3.4% = _____
28. 5.9% = _____

29. 49% = _____
30. 4% = _____
31. 2.1% = _____
32. 3.21% = _____

33. 800% = _____
34. 123% = _____
35. 91% = _____
36. 7.1% = _____

37. 2.8% = _____
38. 900% = _____
39. 147% = _____
40. 117% = _____

41. 7.8% = _____
42. .37% = _____
43. 1100% = _____
44. 912% = _____

45. 24% = _____
46. 8.5% = _____
47. .46% = _____
48. 2300% = _____

49. 721% = _____
50. 15% = _____
51. 9.2% = _____
52. 11.86% = _____

53. 700% = _____
54. 250% = _____
55. 5% = _____
56. 38% = _____

57. 9.9% = _____
58. .55% = _____
59. 2900% = _____
60. 160% = _____

61. 45% = _____
62. 6% = _____
63. 1.3% = _____
64. .092% = _____

65. $\frac{1}{4}$% = _____
66. $\frac{3}{4}$% = _____
67. $\frac{1}{8}$% = _____
68. $\frac{1}{2}$% = _____

69. $9\frac{1}{2}$% = _____
70. $16\frac{1}{4}$% = _____
71. $11\frac{3}{4}$% = _____
72. $11\frac{1}{4}$% = _____

WRITING DECIMALS AS PERCENTS

To convert a decimal to a percent, move the decimal point two places to the right.

EXAMPLES   .29 = 29%   3.6 = 360%   .003 = .3%
            .08 = 8%   .219 = 21.9%   7 = 700%

Convert each decimal to a percent.

1. 3.6 = _____   2. .01 = _____   3. .85 = _____   4. .124 = _____

5. 38 = _____   6. .9 = _____   7. .31 = _____   8. .029 = _____

9. 6.32 = _____   10. 7 = _____   11. .0035 = _____   12. .6 = _____

13. .626 = _____   14. .97 = _____   15. .06 = _____   16. 13 = _____

17. .65 = _____   18. .28 = _____   19. .004 = _____   20. .77 = _____

21. 6 = _____   22. 5.01 = _____   23. .51 = _____   24. 7.2 = _____

25. .464 = _____   26. .813 = _____   27. 8.25 = _____   28. .09 = _____

29. 7.3 = _____   30. 21 = _____   31. .5 = _____   32. .41 = _____

33. .81 = _____   34. .37 = _____   35. 34 = _____   36. .29 = _____

37. .05 = _____   38. 5.9 = _____   39. .73 = _____   40. 4 = _____

41. .32 = _____   42. 113 = _____   43. .008 = _____   44. 4.92 = _____

45. 90 = _____   46. .005 = _____   47. .44 = _____   48. .0061 = _____

49. .007 = _____   50. .03 = _____   51. .009 = _____   52. .91 = _____

53. .999 = _____   54. .102 = _____   55. 29 = _____   56. .894 = _____

57. .93 = _____   58. .71 = _____   59. .68 = _____   60. 71 = _____

61. .88 = _____   62. 4.1 = _____   63. .07 = _____   64. .84 = _____

65. .70 = _____   66. .155 = _____   67. .731 = _____   68. 8 = _____

69. .131 = _____   70. .4 = _____   71. .45 = _____   72. .411 = _____

WRITING PERCENTS AS FRACTIONS

Percent means hundredths.

EXAMPLES

$37\% = \frac{37}{100}$

$48\% = \frac{48}{100} = \frac{12}{25}$

$250\% = \frac{250}{100} = 2\frac{50}{100} = 2\frac{1}{2}$

$\frac{1}{2}\% = \frac{\frac{1}{2}}{100} = \frac{1}{2} \div 100 = \frac{1}{200}$

$33\frac{1}{3}\% = \frac{100}{3} \div 100 = \frac{100}{300} = \frac{1}{3}$

Write each percent as a fraction.

1. 35% = _____

2. 10% = _____

3. 70% = _____

4. 67% = _____

5. 13% = _____

6. 71% = _____

7. 11% = _____

8. 98% = _____

9. 41% = _____

10. 12% = _____

11. 2% = _____

12. 125% = _____

13. 57% = _____

14. 301% = _____

15. 210% = _____

16. 27% = _____

17. 110% = _____

18. 60% = _____

19. 53% = _____

20. 18% = _____

21. 520% = _____

22. 19% = _____

23. 207% = _____

24. 54% = _____

25. 30% = _____

26. 90% = _____

27. 25% = _____

28. 50% = _____

29. 51% = _____

30. 8% = _____

31. 13% = _____

32. 510% = _____

33. 1% = _____

34. 34% = _____

35. 91% = _____

36. 5% = _____

37. 20% = _____

38. 6% = _____

39. 420% = _____

40. 600% = _____

41. $\frac{1}{4}$% = _____

42. $\frac{3}{4}$% = _____

43. $\frac{1}{8}$% = _____

44. $\frac{4}{5}$% = _____

45. $2\frac{1}{2}$% = _____

46. $16\frac{1}{3}$% = _____

47. $8\frac{1}{4}$% = _____

48. $25\frac{1}{2}$% = _____

49. $6\frac{3}{4}$% = _____

50. $8\frac{1}{3}$% = _____

51. $16\frac{1}{4}$% = _____

52. $8\frac{3}{4}$% = _____

53. 65% = _____

54. $9\frac{1}{2}$% = _____

55. $\frac{5}{6}$% = _____

56. $43\frac{1}{2}$% = _____

57. 450% = _____

58. $12\frac{1}{2}$% = _____

59. 38% = _____

60. 9% = _____

61. 58 % = _____

62. $15\frac{1}{4}$% = _____

63. $\frac{1}{5}$% = _____

64. 1250% = _____

WRITING FRACTIONS AS PERCENTS

First, change the fraction to a decimal by division. Round off to hundredths. Then change the decimal to a percent by moving the decimal point two places to the right.

EXAMPLES

$\frac{1}{4} \rightarrow$ 4)1.00 $\rightarrow$ .25 = 25%

$\frac{2}{3} \rightarrow$ 3)2.000 $\rightarrow$ .666 $\rightarrow$ .67 = 67%
$\underline{1\ 8}$
20
$\underline{18}$
20

Write each fraction as a percent. Round to the nearest percent.

1. $\frac{3}{4}$ 4)3.00   .75 = 75%
   $\underline{28}$
   20
   20

2. $\frac{6}{10}$

3. $\frac{1}{5}$

4. $\frac{1}{2}$

5. $\frac{2}{5}$

6. $\frac{3}{8}$

7. $\frac{63}{100}$

8. $\frac{7}{20}$

9. $\frac{8}{25}$

10. $\frac{37}{50}$

11. $\frac{4}{5}$

12. $\frac{1}{4}$

13. $\frac{5}{4}$

14. $\frac{11}{2}$

15. $\frac{7}{4}$

16. $2\frac{1}{4}$

17. $6\frac{1}{2}$

18. $5\frac{1}{5}$

19. $\frac{1}{6}$

20. $\frac{1}{3}$

21. $\frac{5}{7}$

22. $\frac{3}{11}$

23. $\frac{2}{9}$

24. $\frac{5}{8}$

CONVERSIONS

Fractions are converted to decimals by division. Decimals and percents are converted by moving the decimal point. Decimals are converted to fractions by writing them as fractions with the denominator 100.

EXAMPLES   **Convert $\frac{1}{3}$ to a decimal and to a percent.**

$$3\overline{)1.000} \quad .333 \longrightarrow .33 = 33\%$$

Convert 25% to a decimal and to a fraction.

$$25\% = .25 = \frac{25}{100} = \frac{1}{4}$$

Complete the following tables of fractions, decimals, and percents.
Do your work on another sheet of paper.

	Fraction	Decimal	Percent
1.			1%
3.		.05	
5.			8%
7.	$\frac{1}{10}$		
9.			12.5%
11.		.206	
13.	$\frac{3}{4}$		
15.		.25	
17.			40%
19.	$\frac{1}{2}$		
21.		3.75	

	Fraction	Decimal	Percent
2.	$\frac{4}{9}$		
4.		.9	
6.			.03%
8.	$\frac{8}{3}$		
10.			400%
12.		6.5	
14.			8.07%
16.	$2\frac{1}{4}$		
18.		23	
20.	$\frac{7}{12}$		
22.			65%

PERCENT OF A NUMBER

Change each percent to a decimal. Multiply the decimal by the amount of money. Count off the total number of decimal places in the two factors.

EXAMPLES	Find 20% of $120.	Find 10% of $200.	Find 5% of $5.
	20% = .20	10% = .10	5% = .05
	$120	$200	$5.00
	x .20	x .10	x .05
	$24.00	$20.00	$.2500 ⟶ $0.25

Find 25% of each number.

1. $20 _____   2. $500 _____   3. $10 _____   4. $5 _____

Find 6% of each number.

5. $4 _____   6. $300 _____   7. $40 _____   8. $2 _____

Find 15% of each number.

9. $250 _____   10. $8 _____   11. $30 _____   12. $25 _____

Find 8% of each number.

13. $12 _____   14. $900 _____   15. $40 _____   16. $4 _____

Find 10% of each number.

17. $45 _____   18. $560 _____   19. $2 _____   20. $1250 _____

PERCENT OF A DECIMAL

First, convert the percent to a decimal. Next, multiply the decimals. Then round your answer to the nearest hundredth.

EXAMPLES **Find 20% of 1.28.**

$$20\% = .20$$
$$1.28$$
$$\underline{x\ \ .20}$$
$$.2560 \longrightarrow .26$$

Find $6\frac{1}{2}$% of $14.46.

$$6\frac{1}{2}\% = .065$$
$$\$14.46$$
$$\underline{x\ \ \ .065}$$
$$\$.93990 \longrightarrow \$0.94$$

Find the percentage. Round your answer to the nearest hundredth.

1. 70% of 1.3 = _____

2. 9.8% of .6 = _____

3. 84% of .29 = _____

4. 67% of 2.7 = _____

5. 6% of .6 = _____

6. 5% of 6.01 = _____

7. 53% of 2.5 = _____

8. 10% of .27 = _____

9. $6\frac{1}{4}$% of 6.3 = _____

10. $5\frac{1}{2}$% of 71 = _____

11. $9\frac{3}{4}$% of .82 = _____

12. $10\frac{1}{4}$% of .3 = _____

13. 3.5% of 2.5 = _____

14. 12% of 28 = _____

15. $8\frac{1}{2}$% of 621 = _____

16. 5.7% of 42 = _____

17. 5% of 8.5% = _____

18. $9\frac{1}{2}$% of 9 = _____

FINDING PERCENTS

To find what percent one number is of another, divide. Form a fraction with the number following "of" in the denominator. Convert the fraction to a decimal, and then convert the decimal to a percent.

EXAMPLES

5 is what percent of 12?

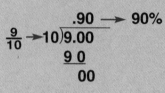

$\frac{5}{12}$ → 12)5.000 = .416 → 42%
4 8
20
12
80
72

What percent is 9 of 10?

$\frac{9}{10}$ → 10)9.00 = .90 → 90%
9 0
00

Find what percent 4 is of 5.

$\frac{4}{5}$ → 5)4.00 = .80 → 80%
4 0
00

---

1. 6 is what percent of 12? _____

2. What percent is 3 of 10? _____

3. Find what percent 9 is of 20. _____

4. 7 is what percent of 8? _____

5. What percent is 1 of 5? _____

6. Find what percent 5 is of 15. _____

7. 11 is what percent of 20? _____

8. What percent is 6 of 100? _____

9. Find what percent 8 is of 100. _____

10. 4 is what percent of 50? _____

11. What percent is 15 of 20? _____

12. Find what percent 10 is of 100. _____

FINDING PERCENTS

3 is what percent of 36?

$$.08\frac{12}{36} = .08\frac{1}{3} = 8\frac{1}{3}\%$$

$$\frac{3}{36} \rightarrow 36\overline{)3.00}$$

$$\underline{2\ 88}$$

$$12$$

1. 21 is _____% of 24

2. 3 is _____% of 18

3. 7 is _____% of 12

4. 2 is _____% of 7

5. 4 is _____% of 36

6. 5 is _____% of 3

7. .3 is _____% of 1.1?

8. 30 is _____% of 48?

9. 8 is _____% of 36?

10. .2 is _____% of 12

11. 12 is _____% of 18

12. 13 is _____% of 6

13. 7 is _____% of 84

14. .5 is _____% of 75

15. 50 is _____% of 7

16. 5 is _____% of 90

17. .3 is _____% of 2.7

18. .36 is _____% of 8.1

FINDING PERCENTS

To find what percent one number is of another, divide. First, form a fraction with the number following "of" in the denominator. Then convert the fraction to a decimal and the decimal to a percent.

EXAMPLES

One day, 3 out of 30 students were absent. What percent were absent?

$$\frac{3}{30} = \frac{1}{10} \longrightarrow 10\overline{)1.00} \qquad \begin{array}{r} .10 \longrightarrow 10\% \\ \underline{1\ 0} \\ 00 \end{array}$$

We drove 150 miles out of 500 on Monday. What percent was that?

$$\frac{150}{500} = \frac{15}{50} = \frac{3}{10} \longrightarrow 10\overline{)3.00} \qquad \begin{array}{r} .30 \longrightarrow 30\% \\ \underline{3\ 0} \\ 00 \end{array}$$

---

1. If you spent $3 out of $15, what percent was that?

Answer: _____

2. If you read 30 of 180 pages, what percent did you read?

Answer: _____

---

3. If you drove 800 miles of a 1000-mile trip, what percent was that?

Answer: _____

4. If a family spent $25 out of $75 for food, what percent was that?

Answer: _____

---

5. If 45 students out of 100 are boys, what percent are boys?

Answer: _____

6. If you spent $75 out of $300, what percent did you have left?   (Careful!)

Answer: _____

---

7. If it rained 3 days out of 8, what was the percent of rainy days?

Answer: _____

8. If 6 out of 36 students were absent, what percent were present?

Answer: _____

120

FINDING A NUMBER WHEN A PERCENT IS KNOWN

When a percentage (number) and percent (%) are given, divide by the percent to find the answer. The percentage is a number which is a given percent of another number. When dividing by a decimal, move the decimal point in both numbers.

EXAMPLES **7 is 20% of what number?**   **30 is 15% of what number?**

$$\frac{7}{.20} \longrightarrow .20\overline{)7.00} \quad \overset{35}{}$$

$$\frac{30}{.15} \longrightarrow 15.\overline{)30.00} \quad \overset{2\ 00}{}$$

1. 4 is 10% of what number? _____

2. 10 is 50% of what number? _____

3. 6 is 30% of what number? _____

4. 12 is 40% of what number? _____

5. 5 is 5% of what number? _____

6. 7 is 14% of what number? _____

7. 20 is 4% of what number? _____

8. 20 is 8% of what number? _____

9. 15 is 3% of what number? _____

10. 100 is 25% of what number? _____

11. 7 is 10% of what number? _____

12. 36 is 50% of what number? _____

13. 11 is 5% of what number? _____

14. 35 is 7% of what number? _____

15. 24 is 10% of what number? _____

16. 110 is 5% of what number? _____

17. 56 is 50% of what number? _____

18. 121 is 25% of what number? _____

FINDING A NUMBER WHEN A PERCENT IS KNOWN

REMEMBER: Read carefully to identify the percentage. To find the number, divide the percentage by the percent.

EXAMPLES

If you went 12 miles and this is 10% of your total trip, how long is the trip?

$$\frac{12}{.10} \longrightarrow .10\overline{)12.00}^{\ 1\ 20}$$

If you spent $3 and this was 5% of your total savings, how much was your total savings?

$$\frac{3}{.05} \longrightarrow .05\overline{)\$3.00}^{\ \$60}$$

---

1. If 2 hours is 25% of the time a job will take, how long will it take to do the whole job?

Answer: _____

2. If $7.50 is 20% of your money, how much money do you have?

Answer: _____

---

3. If 18 miles is 50% of a total distance, what is the distance?

Answer: _____

4. If $14 is 7% of the total cost of a table, how much does the table cost?

Answer: _____

---

5. If $600 is 15% of the cost of a car, what is the cost of the car?

Answer: _____

6. If 2 hours is 10% of the time a trip will take, how long will it take?

Answer: _____

---

7. If 300 miles is 75% of the distance between two cities, what is the distance?

Answer: _____

8. If $6 is 8% of a bill you must pay, how much is the bill?

Answer: _____

---

9. If 10 minutes is 25% of the time for a test, what is the total time?

Answer: _____

10. If $42 is 21% of a total cost, what is the total cost?

Answer: _____

PERCENT OF A NUMBER: SALES TAX

EXAMPLE	Tax rate 3%	$5.00	$5.00
	Amount $5.00	x .03	+ .15
	Tax _____?_____	$.1500 $0.15 Tax	$5.15 Total
	Total _____?_____		

Find the tax and the total amount you would have to pay on the given amount at the given tax rate. Round to the nearest cent.

1. $4.00 at 3%
Tax _____
Total _____

2. $9.00 at 4%
Tax _____
Total _____

3. $10.00 at 5%
Tax _____
Total _____

4. $5.50 at 6%
Tax _____
Total _____

5. $45.00 at 3%
Tax _____
Total _____

6. $37.00 at 4%
Tax _____
Total _____

7. $75.40 at 5%
Tax _____
Total _____

8. $25.90 at 6%
Tax _____
Total _____

9. $23.50 at 6%
Tax _____
Total _____

10. $35.60 at 7%
Tax _____
Total _____

11. $65.70 at 8%
Tax _____
Total _____

12. $39.00 at 5%
Tax _____
Total _____

13. $105.00 at 6%
Tax _____
Total _____

14. $450.00 at 7%
Tax _____
Total _____

15. $234.00 at 8%
Tax _____
Total _____

16. $450.00 at 5%
Tax _____
Total _____

17. $45.20 at $8\frac{1}{2}$%
Tax _____
Total _____

18. $54.24 at $6\frac{1}{2}$%
Tax _____
Total _____

19. $350.80 at $3\frac{1}{2}$%
Tax _____
Total _____

20. $309.05 at $4\frac{1}{2}$%
Tax _____
Total _____

PERCENT OF A NUMBER: DISCOUNT

REMEMBER: A discount is a percent reduction in selling price. To find the new selling price, subtract the discount.

EXAMPLES

If a $22 item is sold at a 15% discount, what is the new selling price?

```
  $22          $22.00   price
x .15         – 3.30    discount
  1 10        $18.70    new price
  2 2
$3.30
```

Find the discount and new price for an item listed at $7.50 with a 20% discount.

```
   $7.50        $7.50   price
x    .20       – 1.50   discount
$1.50 00       $6.00    new price
```

Find the amount of the discount and the new price for each item at the given price and percent discount.

1. $60 at 25% discount _____

2. $400 at 25% discount _____

3. $12.50 at 15% discount _____

4. $45 at 10% discount _____

5. $650 at 5% discount _____

6. $100 at 15% discount _____

7. $9.50 at 30% discount _____

8. $355 at 35% discount _____

9. $150 at 20% discount _____

10. $750 at 12% discount _____

124

PERCENT OF A NUMBER: SALE PRICE

REMEMBER: A sale means that the price of an item is lowered by a certain percentage. After you find the percentage, subtract it from the original selling price. Look at your answer to see if it seems reasonable.

EXAMPLES

A chair selling for $210 was marked "$20% off." What is the new selling price?

$210		$210	price
x .20		– 42	percentage off
$42.00		$168	new selling price

A shirt with a $12 selling price is reduced 30%. What is the new selling price?

$12		$12.00	price
x .30		– 3.60	percentage off
$3.60		$8.40	new price

Find the new selling price for the following items on sale.

1. A $25 lamp at a 10% sale.

   Answer: _____

2. A bed selling for $230 at a 25% sale

   Answer: _____

3. A car selling for $3600 on sale marked "15% off"

   Answer: _____

4. A $35 pair of shoes at a 20% sale

   Answer: _____

5. A $400 moped at a 15% sale

   Answer: _____

6. A jacket selling for $65 at a 20% sale

   Answer: _____

7. A $125 coat at a 10% sale

   Answer: _____

8. A television set selling for $380 at a sale marked "30% off"

   Answer: _____

9. A computer printer selling for $450 at a 25% sale

   Answer: _____

10. A calculator selling for $13.50 at a 15% sale

    Answer: _____

PERCENT OF A NUMBER: PERCENT INCREASE

For a percent increase, find the percentage and then add it. Read each problem carefully. Look at each answer to see if it is reasonable.

EXAMPLES

The Smith family traveled 200 miles one day and the next day went 25% farther. How far did they go on the second day?

```
  200           200
x .25          + 50
10 00           250 miles
40 0
50.00
```

The Jacksons saved $1200 one year. The next year they saved 15% more. How much did they save the second year?

```
  $1200        $1200     first year
x   .15        + 180     increase
  60 00        $1380     second year
 120 0
$180.00
```

1. A train traveled 900 miles one day and went 10% farther the next. How far did it go on the second day?

   Answer: _____

2. A company earned $300,000 one year and 12% more the following year. How much did it earn the second year?

   Answer: _____

3. Chris ran 4 miles one day and 25% farther the next day. How far did Chris run the next day?

   Answer: _____

4. Robin's car gets 25 miles per gallon. Pat's car gets 20% more. How much does Pat's car get?

   Answer: _____

5. Last year the Tildens spent $76 a week for groceries. This year they spent 8% more. How much did they spend this year?

   Answer: _____

6. A house that was sold for $45,000 three years ago is now worth 22% more. How much is it worth now?

   Answer: _____

COMPUTING INTEREST

Interest is calculated on the basis of percent. Change the percent to a decimal, and then multiply to find the interest.

EXAMPLES

Find the interest on $200 at 6%.

$6\% \longrightarrow .06$

$$\begin{array}{r} \$200 \\ \times\ .06 \\ \hline \$12.00 \end{array}$$

Find the interest on $350 at 7%.

$7\% \longrightarrow .07$

$$\begin{array}{r} \$350 \\ \times\ .07 \\ \hline \$24.50 \end{array}$$

Find the interest on each given amount of money at the given rate.

1. $50 at 8% _____

2. $100 at 18% _____

3. $400 at 10% _____

4. $1000 at 5% _____

5. $450 at 7% _____

6. $35 at 5% _____

7. $600 at 11% _____

8. $150 at 18% _____

9. $2500 at 12% _____

10. $5500 at 6% _____

11. $65 at 10% _____

12. $4000 at 12% _____

13. $40 at 8% _____

14. $10,000 at 7% _____

15. $670 at 6% _____

USING INTEREST TO FIND THE PRINCIPAL

REMEMBER: Principal is the money saved or borrowed on which interest is paid. When you know the interest and the rate, divide the interest by the rate to find the principal.

EXAMPLES

Find the principal if interest is $5 at a 10% rate for 1 year.

$$10\% \longrightarrow .10$$

$$\begin{array}{r} \$50 \\ .10\overline{)\$5.00} \end{array}$$

Find the principal if interest is $15 at a 5% rate for 1 year.

$$5\% \longrightarrow .05$$

$$\begin{array}{r} \$300 \\ .05\overline{)\$15.00} \end{array}$$

Find the principal in each exercise. The interest amount and the rate of interest for 1 year are given.

1. $18 at 6% _____

2. $10 at 5% _____

3. $4 at 8% _____

4. $45 at 18% _____

5. $20 at 10% _____

6. $3 at 12% _____

7. $630 at 9% _____

8. $25 at 5% _____

9. $30 at 6% _____

10. $63 at 7% _____

11. $2 at 8% _____

12. $450 at 18% _____

13. $21 at 3% _____

14. $5.46 at 6% _____

15. $5.95 at 7% _____

CALCULATING DOWN PAYMENTS

Down payment is a part of the total cost of an item. It is paid at the time of purchase. The rest is paid month by month. Down payment is a percentage of the total cost. Calculate down payment by converting the percent to a decimal and multiplying the total cost.

EXAMPLES

What is the down payment for a car costing $6000 with a 20% down payment?

$$20\% \longrightarrow .20$$

$$
\begin{array}{r}
\$6000 \\
\times \quad .20 \\
\hline
\$1200.00
\end{array}
$$

Find the amount of a 15% down payment on a $40,000 house.

$$15\% \longrightarrow .15$$

$$
\begin{array}{r}
\$40,000 \\
\times \quad .15 \\
\hline
2000\ 00 \\
4000\ 0 \\
\hline
\$6000.00
\end{array}
$$

---

1. Find the amount of a 10% down payment on a $6000 car.

   Answer: _____

2. Find the amount of a 12% down payment on a $140,000 house.

   Answer: _____

---

3. Find the amount of a 25% down payment on a $8500 car.

   Answer: _____

4. Find the amount of a 30% down payment on a $95,000 house.

   Answer: _____

---

5. Find the amount of a 40% down payment on a $4000 car.

   Answer: _____

6. Find the amount of a 10% down payment on a $65,000 house.

   Answer: _____

---

7. Find the amount of a 50% down payment on a $6500 car.

   Answer: _____

8. Find the amount of a 15% down payment on a $88,000 home.

   Answer: _____

BUYING ON CREDIT

E X A M P L E **Anne and Fred Williams used their credit card to buy a new refrigerator that cost $700. They paid for the refrigerator in monthly payments of $70 each, beginning in July. They also paid a finance charge of 1% each month on the unpaid balance. Their July balance was found as follows:**

$700.00	Beginning balance	$630.00	New amount
− 70.00	July payment	+ 6.30	Finance charge
$630.00	New amount	$636.00	New balance
x .01	Monthly rate		
$6.3000			

**Complete the table below to find the Williams' balance after each month.
Round off the amount of the finance charge to the next higher cent.
Do your work on a separate sheet of paper.**

	Month	Balance	Payment	New Amount	1% Finance Charge	Balance
	July	$700.00	$70.00	$630	$6.30	$636.30
1.	August	$636.30	$70.00	_____	_____	_____
2.	September	_____	$70.00	_____	_____	_____
3.	October	_____	$70.00	_____	_____	_____
4.	November	_____	$70.00	_____	_____	_____
5.	December	_____	$70.00	_____	_____	_____
6.	January	_____	$70.00	_____	_____	_____
7.	February	_____	$70.00	_____	_____	_____
8.	March	_____	$70.00	_____	_____	_____
9.	April	_____	$70.00	_____	_____	_____
10.	May	_____	_____	$0		

PROBLEMS INVOLVING PERCENTS

If a chair priced at $240 has a 25% discount, what is the selling price?

$240	$240
x .25	- 60
12 00	$180 **Answer**
48 0	
$60.00	

If you spent $18 out of $90, what percent did you spend?

$$\frac{18}{90} \longrightarrow \quad \overset{.20 \longrightarrow 20\%}{\$90\overline{)\$18.00}}$$

---

1. What is the amount of a 7% tax on $43?

Answer: _____

2. What is the new selling price of a $46 table at a 15% sale?

Answer: _____

---

3. Mrs. Smith bought a house for $30,000 and sold it for 20% more. What was the selling price?

Answer: _____

4. If 5 out of 40 were absent, what percent were absent?

Answer: _____

---

5. If 24 miles is 50% of a distance, what is the distance?

Answer: _____

6. If $25 is 10% of a total cost, what is the total cost?

Answer: _____

---

7. Find the interest on $60 at 6% for 1 year.

Answer: _____

8. Find the amount of a 35% down payment on a $40,000 home.

Answer: _____

PROBLEMS INVOLVING PERCENTS

If $2500 is 5% of the cost of a house, find the cost.

$5\% \longrightarrow .05 \quad .05)\overline{2500.00} \quad \begin{array}{c} 500.00 \\ \hline \end{array} \longrightarrow \$50,000$

Find the interest on $500 at 4% for 1 year.

$$4\% \longrightarrow .04 \qquad \begin{array}{r} \$500 \\ \times \quad .04 \\ \hline \$20.00 \end{array}$$

---

1. If the tax on a $4500 car was $360, find the percent tax.

Answer: _____

2. If on a 10-hour trip you plan to rest $1\frac{1}{2}$ hours, what percent is this?

Answer: _____

---

3. If a house costs 60,000 and you pay a 12% down payment, how much is this?

Answer: _____

4. If you must pay $3000 in income tax and this is 15% of you total income, what is your income?

Answer: _____

---

5. Find the interest on $12,000 for 1 year at 5%.

Answer: _____

6. Find the principal if the interest is $33 at 6% for 1 year.

Answer: _____

---

7. Find the amount of a 10% down payment on an $86,00 house.

Answer: _____

8. If 1 hour is 5% of the time for a trip, how long will the trip take?

Answer: _____

PROBLEMS INVOLVING PERCENTS

If 1 hour is 5% of the time needed to do a job, how long will it take to do the job?

$$5\% \longrightarrow .05 \qquad .05\overline{)1.00} \quad \overset{20}{} \text{ hours}$$

If $400 is 8% of the cost of a car, how much does the car cost?

$$8\% \longrightarrow .08 \qquad .08\overline{)400.00} \quad \overset{5000}{} \longrightarrow \$5000$$

---

1. What is a 7% tax on $450?

Answer: _____

2. What is 25% of 2 days in hours?

Answer: _____

---

3. If an 8% sales tax is $12, what is the total amount?

Answer: _____

4. Find the amount of a 12% down payment on a $24,000 house.

Answer: _____

---

5. What is 50% of 3 hours?

Answer: _____

6. If $1\frac{1}{2}$ hours is 25% of the time needed to do a job, how long will the job take?

Answer: _____

---

7. What percent of an hour is 45 minutes?

Answer: _____

8. What percent of 8 days is 1 day?

Answer: _____

PROBLEMS INVOLVING PERCENTS

If 10% of a trip takes $2\frac{1}{2}$ hours, how long will the trip take?

$$10\% \longrightarrow .10 \qquad .10\overline{)2.50}^{\underline{25}\ \text{hours}}$$

What will be a 12% down payment on a $35,000 house?

$$12\% \longrightarrow .12 \qquad \begin{array}{r} \$35,000 \\ \times \quad .12 \\ \hline 700\ 00 \\ 3500\ 0 \\ \hline \$4200.00 \end{array}$$

---

1. If you make a $1000 down payment on a $4500 car, what percent is your down payment?

Answer: _____

2. What percent of 10 hours is 30 minutes?

Answer: _____

---

3. What is the amount of a 4% tax on $12.50?

Answer: _____

4. What will be a 25% down payment on a $15,000 house?

Answer: _____

---

5. If a chair costing $90 is on sale at a 15% discount, what will be the new selling price?

Answer: _____

6. If the price of a $42,000 house is increased by 20%, what is the new price?

Answer: _____

---

7. If prices go up 6%, what will be the new cost of a $35 item?

Answer: _____

8. What will be the cost of a $12 shirt at a 25% sale?

Answer: _____

C H A L L E N G E

For percents that are close to 25%, 50%, and 75%, you can use $\frac{1}{4}$, $\frac{1}{2}$, and $\frac{3}{4}$ to estimate.

EXAMPLE

In an election, Mr. Palumbo received 28% of the 12,219 votes in his district. About how many votes did he receive?

He received a little more than 25%, or $\frac{1}{4}$, of the votes.

$\frac{1}{4}$ of 12,219 is a little more than 3000.

So, Mr. Palumbo received a little more than 3000 votes.

1. In the same election, Ms. Velez received 7103 of the 12,219 votes. Was that more or less than the 50% needed to win the election?

   Answer: _____

2. A shirt that usually sells for $19.95 is on sale at a 70% discount. Does DeShawn need a $10 bill or a $20 bill to buy the shirt?

   Answer: _____

3. On a test, Otto scored 68%. He took the test again. This time, he had 31 out of 40 correct. Did he do better or worse the second time?

   Answer: _____

4. The Ramdath family bought their home for $120,000. Since then, it has increased in value by 48%. About how much is their home worth now?

   Answer: _____

5. Mr. and Mrs. Baum try to spend no more than $\frac{1}{4}$ of their family income on rent. They earn $3182 per month. Which of the apartments listed at the right can they afford?

   Answer: _____

   <u>Apple Tree Manor Apartments</u>

Studio Apartment	$350/month
1-Bedroom	$450/month
2-Bedroom	$600/month
3-Bedroom	$725/month

6. Ms. Unruh is buying a new car for $23,785. She is required to put 25% down. About how much is that?

Answer: _____

7. One poll showed that 68% of the people surveyed said that there is too much violence on television. If 600 people were surveyed, about how many said there was too much violence – 300, 400, or 500?

Answer: _____

8. The Humongo Cheeseburger Deluxe Meal has 77% of an adult's maximum daily allowance of dietary fat. If an adult is allowed about 80 grams of fat per day, about how many grams does the meal have?

Answer: _____

9. A poll showed that 543 people thought the mayor was doing a good job and 489 thought he was doing a bad job. Was the mayor's popularity rating above 50%?

Answer: _____

10. ABC Hardware sells widgets for $125. XYZ Hardware usually sells them for $159, but they are on sale for 25% off. Which store has the better buy?

Answer: _____

11. Mr. and Mrs. Green have a coupon that gives 50% off the second dinner they buy. If each of their dinners is $21.95, will $30.00 be enough to pay for both dinners?

Answer: _____

12. On one test, Esteban got 65 correct out of 81 problems. Ming got 67% correct. Who got the higher score?

Answer: _____

13. Ms. Stevens pays 21% of her income in federal income tax and 5% in state income tax. If she earns $60,000 annually, about how much does she pay in income tax?

Answer: _____

POSTTEST

Convert the following percents to decimals.

1. 35% = _____

2. 414% = _____

3. $3\frac{1}{2}$% = _____

4. 9.8% = _____

5. 6% = _____

6. 12% = _____

Convert the following percents to fractions. Then reduce.

7. 36% = _____

8. 150% = _____

9. $4\frac{1}{2}$% = _____

Convert the following to percents.

10. .54 = _____

11. 48 = _____

12. .713 = _____

13. 6.1 = _____

14. .02 = _____

15. .9 = _____

Convert the following to percents.

16. $\frac{3}{4}$ = _____

17. $\frac{5}{6}$ = _____

18. $\frac{5}{8}$ = _____

19. 4 = _____

20. Find 75% of 180. _____

21. Find 3.8% of $2.40. _____

22. 8 is what percent of 20? _____

23. 7.2 is what percent of 12? _____

24. If you spent $15 out of $20, what percent did you spend?

Answer: _____

25. 18 is 3% of what number?

Answer: _____

26. If a $60 item is sold at a 15% discount, what is the new selling price?

Answer: _____

27. What would be the total price of a $96.50 item with a 6% sales tax?

Answer: _____

28. What is the price of a table that usually sells for $180, but is marked "25% off"?

Answer: _____

29. A house worth $55,000 2 years ago is now worth 24% more. What is it worth now?

Answer: _____

30. If 45 miles is 30% of a total distance, what is the distance?

Answer: _____

31. Find the interest on $1500 at 6% for 1 year.

Answer: _____

32. Find the principal if the interest at 12% is $84 for 1 year.

Answer: _____

33. If a car costs $5900 and you must make a 30% down payment, what is the down payment?

Answer: _____

CUMULATIVE REVIEW

Add.

1. 216
   + 85

2. 24.61 + 33.2 = _____

3. $67.94
   +29.86

4. $\frac{3}{5} + \frac{2}{3}$ = _____

Subtract

5.   281
   − 138

6. $9\frac{3}{4}$
  − $2\frac{1}{2}$

7. 6
 − $\frac{1}{3}$

8. $100.00
  − 29.56

Multiply

9. 6037
  x 25

10. $45.67
   x   9

11. $1\frac{5}{6}$ x 4 = _____

12. 3.68
  x .2

Divide.

13. $8\overline{)19,656}$

14. $12\overline{)303.60}$

15. $6 \div \frac{2}{3}$ = _____

16. Subtract forty-nine dollars from one hundred eleven dollars.

Answer: _____

17. If a hat costs $12.50 and shoes cost $35.90, how much do 1 hat and 2 pairs of shoes cost?

Answer: _____

18. If you paid $13.05 for 9 gallons of gas, how much was the cost per gallon?

Answer: _____

19. If you had $25 and spent one-fourth of it, how much money would you have left?

Answer: _____

20. How many hours are there in $\frac{2}{3}$ day?

Answer: _____

21. What is one-tenth of $200?

Answer: _____

22. Raise to the power: 4^3 = _____

23. Reduce $\frac{12}{48}$ to lowest terms. _____

24. Round 6358 to the nearest hundred. _____

25. Write the number two thousand thirty. _____

Convert the following percents to decimals.

26. 70% = _____   **27.** 1.25% = _____   **28.** 650% = _____   **29.** 3% = _____

Convert the following percents to fractions.

30. 75% = _____   **31.** 630% = _____   **32.** 10% = _____   **33.** 5% = _____

34. Find 25% of 180.

Answer: _____

35. Find 20% of $80.00

Answer: _____

36. If a $55 item is sold at 10% discount, find the new selling price.

Answer: _____

37. What would be the total on a charge of $450.80 with 4% sales tax?

Answer: _____

38. Find the price of a table selling for $175, but marked "10% off."

Answer: _____

39. A car sold for $4000 two years ago is now worth 20% less. What is it worth now?

Answer: _____

Convert the following fractions to decimals and then to percents.

40. $\frac{1}{5}$ = _____   **41.** $\frac{1}{25}$ = _____   **42.** $\frac{3}{10}$ = _____

43. 6 is what percent of 10?

Answer: _____

44. What percent is 35 of 50?

Answer: _____

45. If you spent $6 out of $30, what percent did you spend?

Answer: _____

46. 25 is 10% of what number?

Answer: _____

47. If 45 miles is 20% of a total distance, what is the distance?

Answer: _____

48. Find the interest on $900 at 4% for 1 year.

Answer: _____

49. Find the principal if the interest is $16 at 5% for 1 year.

Answer: _____

50. If a house costs $55,000 and you must make a 15% down payment, what is the down payment?

Answer: _____

140

PRETEST

Estimate the length of the line below in inches and also in centimeters.

_____   **1.** _____ in.   **2.** _____ cm

3. The length of a golf club is about (check one)

    1 meter _____   1 kilometer _____   1 millimeter _____

Convert each of the following lengths to meters by multiplying or dividing.

   4. 8 kilometers = _____ meters   **5.** 528 centimeters = _____ m

Convert each of the following lengths to feet.

   6. 6 yd = _____ ft   **7.** 72 in. = _____ ft   **8.** 2 mi = _____ ft

Perform the operations indicated.

   9.   1' 6"   **10.**   4' 7   **11.**   9' 5"   **12.**   4' 7"
      +2' 3"   +2' 6"   − 3' 9"   x 3

                                                  6"

                      4"

13. Find the perimeter of the rectangle at the right.
Write the answer in feet and inches.

                Perimeter _____

14. A pair of shoes weighs about (check one)   1 gram _____   1 kilogram _____

Convert each of the following.

   15. 7.68 kg = _____ g   **16.** 48 grams = _____ kilograms

   17. 6 lb = _____ oz   **18.** 80 oz = _____ lb   **19.** 4.5 tons = _____ lb

20. A glass of milk contains about (check one)   200 mL _____   200L _____

Convert each measure to quarts.

21. 7 gallons = _____ quarts

22. 24 pints = _____ quarts

Convert each of the following to minutes.

23. 300 sec = _____ min

24. 9 hours = _____ min

Perform the operations indicated.

25.   6 hr 14 min
        + 3 hr 52 min

26.   5 min 30 sec
        − 2 min 19 sec

27.   6 days 9 hr
        − 1 days 15 hr

Find the elapsed time in each exercise.

28. 10:20 A.M. to 1:40 P.M.

29. 9 A.M. Tues. to 11 A.M. Wed.

Answer: _____

Answer: _____

30. On a good day for swimming, the air temperature might be

(check one)   25°F _____   25°C _____

31. Lois is 69 inches tall. How tall is she in feet and inches?

Answer: _____

32. What part of a week is 2 days?

Answer: _____

33. .83 L = _____ mL

142

ESTIMATING AND MEASURING LENGTH

A centimeter (cm) is about the width of a fingernail. A millimeter (mm) is about the thickness of a dime.

EXAMPLE **Estimate the length of this bar in inches, centimeters, and millimeters. Then use a ruler to find the actual length.**

	Estimate	Measure
Inches	3	$2\frac{11}{16}$
Centimeters	6	6.8
Millimeters	60	68

**Estimate the length of each bar in inches, centimeters, and millimeters.
Then use a ruler to find the actual measurements.**

	Estimate			Measurement		
	in.	cm	mm	in.	cm	mm
1.	___	___	___	___	___	___
2.	___	___	___	___	___	___
3.	___	___	___	___	___	___
4.	___	___	___	___	___	___
5.	___	___	___	___	___	___
6.	___	___	___	___	___	___
7.	___	___	___	___	___	___
8.	___	___	___	___	___	___
9.	___	___	___	___	___	___
10.	___	___	___	___	___	___
11.	___	___	___	___	___	___
12.	___	___	___	___	___	___

CHOOSING SUITABLE UNITS OF LENGTH

REMEMBER: A centimeter is small, about the width of a fingernail. A meter is equal to 100 centimeters. A person's waist is about 1 meter high. A kilometer is equal to 1000 meters. It takes about 12 minutes to walk a kilometer.

EXAMPLES **Would you use centimeters, meters, or kilometers to measure the following?**

The length of a pencil

cm (centimeters)

The height of a building

m (meters)

The distance between cities

km (kilometers)

Indicate whether you would use centimeters, meters, or kilometers to measure each of the following.

1. The height of a table _____

2. Distances for short races _____

3. The length of a car trip _____

4. The height of a picture _____

5. The width of a table _____

6. Distances on a map _____

7. The length of a belt _____

8. The length of a soccer field _____

9. The height of a person _____

10. The distance around the world _____

11. Distances on a train schedule _____

12. The height of a house _____

13. The height of a flag pole _____

14. The length of a pair of skis _____

15. The width of a television screen _____

16. The distance to the moon _____

17. The height of a salt shaker _____

18. The length of a piece of paper _____

19. Distance from New York to Chicago _____

20. The length of a city block _____

21. The height of a lamp _____

22. Distance across a state _____

Circle the correct measure for each of the following.

23. The height of a man: 180 cm   180 m   180 km

24. Distance from Los Angeles to San Francisco: 690 cm   690 m   690 km

25. Length of a new pencil: 18 cm   18 m   5000 km

26. Long-distance run: 5000 cm   5000 m   5000 km

CHOOSING SUITABLE UNITS OF LENGTH

The length of a forefinger is about 7 centimeters. A telephone pole is about 6 meters high. You could walk a kilometer in about 12 minutes.

EXAMPLES **Select the most logical answer, and then write it on the line next to the distance or size it goes with.**

Length of this page	__28 cm__	**960 kilometers**
Distance from Denver to Kansas City	__960 km__	**30 meters**
Height of a 10-story building	__30m__	**28 centimeters**

Read each size or distance. Then choose the measurement from the column on the right that should go with it.

1. Height of a tree _____ 1 centimeter

2. Length of a pencil _____ 17 centimeters

3. Distance from New York to California _____ 1 meter

4. Width of a fingernail _____ 8 meters

5. Length of a golf club _____ 4800 kilometers

6. Height of a person _____ 180 centimeters

7. Height of a flagpole _____ 7 meters

8. Speed Limit (per hour) _____ 100 meters

9. Distance around the world _____ 81 kilometers

10. Length of an Olympic race _____ 40,000 kilometers

11. Length of a shoe _____ 28 centimeters

12. Length of a long-distance race _____ 205 centimeters

13. Length of skis _____ 60 meters

14. Distance from Miami to Houston _____ 25 kilometers

15. Height of a 20-story building _____ 1900 kilometers

METRIC-METRIC CONVERSIONS OF LENGTH

REMEMBER: Conversion in the metric system means merely moving the decimal point. When going from a larger to a smaller measure you multiply. So you move the decimal point to the right. When going from a smaller to a larger measure you divide. So you move the decimal point to the left. If there is no decimal point, you put one at the end. Add zeros as needed.

1 kilometer = 1000 meters
1 meter = 100 centimeters

E X A M P L E S   **Convert to the indicated unit.**

12 meters to centimeters   __1200 cm__   **(Multiply by 100.)**

8 meters to kilometers   __.008 km__   **(Divide by 1000.)**

Convert to the indicated unit.

1. 10 meters to centimeters _____

2. 6.7 m to cm _____

3. 100 centimeters to meters _____

4. 56.3 km to m _____

5. 10 kilometers to meters _____

6. 9.60 m to cm _____

7. 2.5 meters to centimeters _____

8. 34,000 m to km _____

9. 2000 meters to centimeters _____

10. 21,000 cm to m _____

11. 250 centimeters to meters _____

12. 57.3 m to cm _____

13. 6.5 kilometers to meters _____

14. 91,000 m to km _____

15. 1000 centimeters to meters _____

16. 52.21 km to m _____

17. 10.8 meters to centimeters _____

18. 3800 m to km _____

19. 31 kilometers to meters _____

20. 2200 cm to m _____

21. 3500 centimeters to meters _____

22. 13 km to m _____

23. 6500 meters to kilometers _____

24. 8.01 m to cm _____

25. 12.25 kilometers to meters _____

26. 200 cm to m _____

146

CONVERSIONS OF CUSTOMARY LENGTH

REMEMBER: When changing to a smaller unit, multiply. When changing to a larger unit, divide. Use the table below.

12 inches (in.) = 1 foot (ft)
36 inches = 1 yard (yd)
3 feet = 1 yard

5280 feet = 1 mile (mi)
1760 yards = 1 mile

EXAMPLES **Convert both 48 inches and 3 miles to feet.**

48 inches ÷ 12 = 4 feet   3 miles x 5280 = 15,840 feet

Convert to the indicated unit.

1. 6 ft = _____ in.

2. 96 in. = _____ ft

3. 10,560 ft = _____ mi

4. 9 mi = _____ ft

5. 72 in. = _____ yd

6. 8 yd = _____ in.

7. 5 yd = _____ ft

8. 21 ft = _____ yd

9. 6 mi = _____ yd

10. 5280 yd = _____ mi

11. 132 in. = _____ ft

12. 5 mi = _____ ft

13. 9 yd = _____ in.

14. 10,560 yd = _____ mi

15. 51 ft = _____ yd

16. 9 ft = _____ in.

17. 15,840 ft = _____ mi

18. 192 in. = _____ yd

19. $13\frac{1}{2}$ mi = _____ yd

20. 78 in. = _____ ft

21. $5\frac{1}{3}$ yd = _____ in.

22. 38 ft = _____ in.

23. 16,720 yd = _____ mi

24. $4\frac{1}{2}$ yd = _____ in.

ADDING UNITS OF LENGTH

Twelve inches equals one foot (12" = 1'). You should rename the inches when the sum has more than 12 inches.

EXAMPLES

3' 6"	5' 4"
+2' 4"	+6' 9"
5' 10"	11' 13" = 12'1"

Since 13" is more that 12", write 13" as 1'1" and carry 1'.

Add. Rename and carry when necessary.

1. 5' 5"
　　+6' 2"

2. 4' 8"
　　+6' 1"

3. 3' 3"
　　+2' 5"

4. 10' 4"
　　+21' 7"

5. 23' 8"
　　+17' 3"

6. 5' 7"
　　+2' 6"

7. 6' 8"
　　+4' 6"

8. 9' 5"
　　+6' 9"

9. 12' 6"
　　+ 9' 5"

10. 13' 9"
　　+ 9' 10"

11. 24' 6"
　　+35' 6"

12. 15' 7"
　　+36' 7"

13. 46' 8"
　　+27' 8"

14. 57' 9"
　　+65' 9"

15. 76' 10"
　　+57' 10"

16. 9' 3"
　　8' 6"
　　+5' 8"

17. 6' 5"
　　6' 0"
　　+3' 7"

18. 12' 4"
　　6' 9"
　　+16' 10"

19. 25' 11"
　　36' 2"
　　+19' 8"

20. 9' 8"
　　45' 7"
　　+26' 2"

21. 3' 11"
　　3' 10"
　　+4' 3"

22. 8' 9"
　　6' 0"
　　+7' 9"

23. 14' 11"
　　16' 8"
　　+22' 9"

24. 32' 10"
　　4' 10"
　　+17' 5"

25. 57' 7"
　　29' 10"
　　+38' 11"

26. Claudia needs 20 feet of lumber for a box she is building. She has 3 pieces. One is 8 feet 5 inches, one is 4 feet 8 inches, and one is 6 feet 10 inches. Add the lengths of the three pieces to find if she has enough lumber.

Answer: _____

SUBTRACTING UNITS OF LENGTH

REMEMBER: If you need to borrow, subtract 1 from the feet and add 12 to the inches.

EXAMPLES

5' 7"	6' 4" → 5' 16"	Since 7" is more than 4", borrow
−2' 3"	−2' 7" → −2' 7"	1' and add 12" to 4".
3' 4"	3' 9"	

Subtract. Borrow when necessary.

1. 8' 7"
    −2' 5"

2. 9' 6"
    −5' 3"

3. 10' 10"
    − 8' 5"

4. 12' 8"
    − 3' 5"

5. 15' 11"
    − 4' 7"

6. 28' 11"
    −13' 8"

7. 47' 10"
    − 9' 2"

8. 76' 10"
    −57' 10"

9. 13' 8"
    − 9' 0"

10. 46' 10"
     − 2' 3"

11. 9' 4"
     −5' 9"

12. 8' 7"
     −3' 8"

13. 16' 8"
     −10' 9"

14. 6' 0"
     −3' 9"

15. 22' 2"
     −17' 10"

16. 19' 6"
     − 7' 11"

17. 10' 3"
     − 7' 9"

18. 6' 2"
     −5' 5"

19. 44' 0"
     −16' 11"

20. 19' 2"
     −11' 8"

21. 51' 3"
     −19' 11"

22. 28' 1"
     − 9' 9"

23. 13' 0"
     −12' 11"

24. 16' 4"
     −   9"

25. 60' 10"
     −19' 11"

MULTIPLYING UNITS OF LENGTH

REMEMBER: Multiply by feet and inches separately. If the inches are more than 12, convert inches to feet by dividing. For example, 56 inches equals 4 feet and 8 inches.

Multiply. Convert inches to feet wherever possible.

1. 3' 2"
   x 3

2. 4' 2"
   x 4

3. 5' 2"
   x 5

4. 10' 3"
   x 3

5. 12' 4"
   x 2

6. 10' 4"
   x 3

7. 13' 3"
   x 4

8. 15' 5"
   x 3

9. 18' 6"
   x 2

10. 20' 5"
    x 4

11. 21' 2"
    x 6

12. 22' 7"
    x 3

13. 25' 4"
    x 4

14. 27' 7"
    x 5

15. 28' 8"
    x 7

16. 30' 9"
    x 5

17. 32' 7"
    x 7

18. 39' 10"
    x 3

19. 34' 11"
    x 2

20. 36' 10"
    x 6

21. 43' 8"
    x 12

22. 45' 11"
    x 10

23. 49' 9"
    x 9

24. 50' 11"
    x 15

25. 54' 7"
    x 6

26. Mario measured the length of his stride as 2 feet 10 inches. How far would he walk in 20 strides?

Answer _____

PERIMETERS

Perimeter means distance around. In a square, each side is the same length. In a rectangle, the widths are the same and the lengths are the same.

EXAMPLES **Find the perimeter of each figure.**

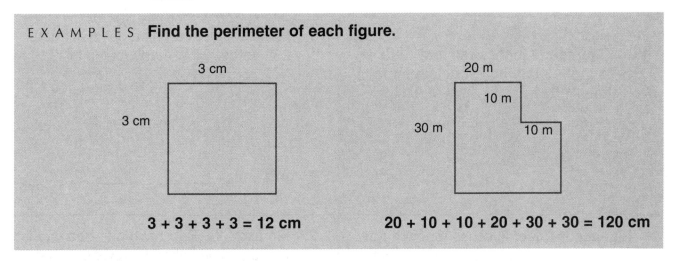

3 + 3 + 3 + 3 = 12 cm   20 + 10 + 10 + 20 + 30 + 30 = 120 cm

Find the perimeter of each figure.

1.

15 km
15 km

2.

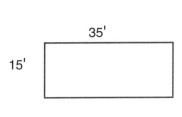
35'
15'

3.

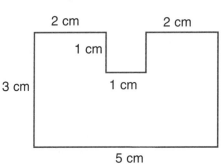
40 m
30 m
10 m
20 m

4.

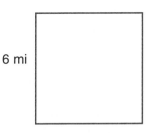
6 mi

5.

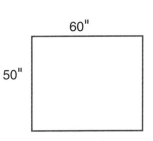
60"
50"

6.

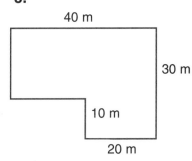
2 cm   2 cm
1 cm
3 cm   1 cm
5 cm

7.

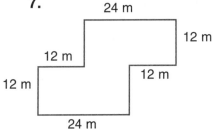
24 m
12 m
12 m
12 m
12 m
24 m

8.

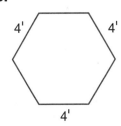
4'   4'
4'

9.

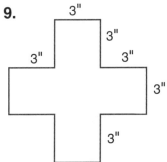
3"
3"
3"   3"
3"
3"

PROBLEMS USING UNITS OF LENGTH

Read each problem carefully. Think about whether to add, subtract, or multiply. Work the problem, and see if your answer makes sense.

EXAMPLE

If Ron is 6 feet 2 inches tall and Anita is 5 feet 5 inches tall, how much taller is Ron than Anita?

6' 2" ⟶ 5' 14"
-5' 5" ⟶ -5' 5"
9"

---

1. Maria's father can hit a golf ball about 280 yards. How many feet is this?

Answer: _____

---

2. Rita needs 2 feet 6 inches for each row of tomatoes in her garden. How much room is needed for 5 rows?

Answer: _____

3. Mr. and Mrs. Barr have a fir tree that is 6 feet 8 inches tall. If it grows 6 inches during the next year, how tall will it be?

Answer: _____

---

4. George can reach 8 feet 3 inches high. How high can he reach when standing 6 feet 11 inches on a ladder?

Answer: _____

5. Route 83 has 4 lanes. Each lane is 11 feet 9 inches wide. How wide is Route 83?

Answer: _____

---

6. Myra is 63 inches tall. How tall is she in feet and inches?

Answer: _____

7. Last winter it snowed 1 foot 5 inches in December, 2 feet 9 inches in January, and 11 inches in February. How much snow fell in all?

Answer: _____

CHOOSING SUITABLE UNITS OF WEIGHT

REMEMBER: A gram is a small amount. A sunflower seed weighs about 1 gram. Some books weigh about 500 grams. A kilogram is equal to 1000 grams. Some babies weigh about 3 kilograms when they are born. Some professional football players weigh more than 110 kilograms.

EXAMPLES **Would you use grams or kilograms for the following?**

A baseball   grams   A child   kilograms

Read each item on the left, and then choose the measure from the column on the right to go with it. Write the measure on the line next to the item.

1. A pencil _____ 60 kilograms

2. A computer _____ 120 grams

3. A 16-year-old person _____ 10 kilograms

4. An apple _____ 15 grams

5. A steak on your plate _____ 5 grams

6. A car _____ 1500 kilograms

7. A big dog _____ 200 grams

8. A nickel _____ 40 kilograms

9. A grown person _____ 1 kilogram

10. A peach _____ 1000 kilograms

11. A hammer _____ 80 kilograms

12. An elephant _____ 100 grams

13. A bowling ball _____ 150 kilograms

14. A tiger _____ 2 grams

15. A pin _____ 7 kilograms

16. A dime _____ .2 grams

METRIC-METRIC CONVERSIONS OF WEIGHT

REMEMBER: 1 kilogram (kg) equals 1000 grams (g). To convert kilograms to grams, multiply by 1000. To convert grams to kilograms, divide by 1000.

EXAMPLES **Convert 8.2 kilograms to grams.** **8200 grams**
Convert 63 grams to kilograms. **.063 kilograms**

Convert to the indicated unit.

1. 47 kilograms = _____ grams

2. 604 grams = _____ kilograms

3. .8 kilograms = _____ grams

4. 58.1 grams = _____ kilograms

5. 128 grams = _____ kilograms

6. .146 kilograms = _____ grams

7. .64 grams = _____ kilograms

8. 9.3 kilograms = _____ grams

9. .06 kilograms = _____ grams

10. 1 gram = _____ kilograms

11. .098 kilograms = _____ grams

12. .07 grams = _____ kilograms

13. 5.2 grams = _____ kilograms

14. 43 kilograms = _____ grams

15. 959 grams = _____ kilograms

16. 6.8 kilograms = _____ grams

17. .54 kilograms = _____ grams

18. 7310 grams = _____ kilograms

19. 23 grams = _____ kilograms

20. .092 kilograms = _____ grams

21. 65 kg = _____ g

22. 219 g = _____ kg

23. .4 kg = _____ g

24. 29.3 = _____ kg

25. 97,000 g = _____ kg

26. 7.31 kg = _____ g

27. .71 g = _____ kg

28. 6.8 kg = _____ g

29. .09 kg = _____ g

30. 70,000 g = _____ kg

31. 38 kg = _____ g

32. .05 g = _____ kg

33. 2.6 g = _____ kg

34. 46 kg = _____ g

35. 453 g = _____ kg

36. 3.5 kg = _____ g

154

CONVERSIONS OF CUSTOMARY WEIGHT

REMEMBER: 16 ounces (oz) equals 1 pound (lb). 2000 pounds equals 1 ton. Multiply when you convert to a smaller unit. Divide when you convert to a larger unit.

EXAMPLES **Convert 48 ounces to pounds.**   **Convert 6 tons to pounds.**

48 ÷ 16 = 3 pounds   6 x 2000 = 12,000 pounds

Convert to the indicated unit.

1. 64 ounces = _____ pounds

2. 3 tons = _____ pounds

3. 10 pounds = _____ ounces

4. 8000 pounds = _____ tons

5. 9 tons = _____ pounds

6. 96 ounces = _____ pounds

7. 14,000 pounds = _____ tons

8. 17 pounds = _____ ounces

9. 128 ounces = _____ pounds

10. 11 tons = _____ pounds

11. 5 lb = _____ oz

12. 6000 lb = _____ tons

13. 19 tons = _____ lb

14. 32 oz = _____ lb

15. 20,000 lb = _____ tons

16. 35 lb = _____ oz

EXAMPLES **Convert 6.5 pounds to ounces.**   **Convert 24 ounces to pounds.**

16 x 6.5 = 104.0   24 ÷ 16 = 1.5

17. 8.5 lb = _____ oz

18. 40 oz = _____ lb

19. 3000 lb = _____ tons

20. 5.5 tons = _____ lb

21. 84 oz = _____ lb

22. 4.75 lb = _____ oz

23. 9.8 tons = _____ lb

24. 4500 lb = _____ tons

25. 600 pounds = _____ tons

26. 74 oz = _____ lb

27. $7\frac{3}{4}$ lb = _____ oz

28. $4\frac{1}{5}$ tons = _____ pounds

PROBLEMS INVOLVING UNITS OF WEIGHT

EXAMPLE

If Jack weighs 70 kilograms and Mary weighs 61 kilograms, how much heavier is Jack than Mary?

$$\begin{array}{r} 70 \\ -61 \\ \hline 9 \text{ kg} \end{array}$$

1. If you bought 2 kilograms of meat and 1.5 kilograms of vegetables, how many kilograms did you buy in all?

   Answer: _____

---

2. If you bought 1.25 kilograms of fish at $7 a kilogram, how much did you have to pay?

   Answer: _____

3. If you bought a pie weighing 1.2 kilograms and divided it among 6 people, how much did each get?

   Answer: _____

---

4. If you bought 10 grams of medicine at $0.75 a gram, how much did you pay?

   Answer: _____

5. If 3 kilograms of meat cost $12, how much does .75 kilogram cost?

   Answer: _____

---

6. If you bought 1.5 kilograms of fruit for $2 a kilogram and half went bad, how much money did you waste?

   Answer: _____

7. If 200 grams of cheese cost $1, how much will 600 grams cost?

   Answer: _____

---

8. If a bridge will hold only 4500 kilograms and a truck weighs 2650 kilograms, what is the heaviest load it can carry across the bridge?

   Answer: _____

9. The shipping charge for certain mail orders is $0.87 per kilogram. How much is the cost for shipping an order weighing 6.9 kilograms?

   Answer: _____

METRIC UNITS OF CAPACITY

REMEMBER: 1 liter of milk is slightly more than 1 quart. 1 milliliter of water is about 2 large drops. Abbreviations are L (liter) and mL (milliliter). 1 liter equals 1000 millimeters. To convert from liters to milliliters, multiply by 1000. To convert from milliliters to liters, divide by 1000.

EXAMPLES

A gallon of gas is about 4.24 liters.

4.24 L = 4240 mL

The capacity of a coffee cup is about 200 milliliters.

200 mL = .2 mL.

Circle the most sensible measure.

1. Glass of water: 24 L   24 mL   240 mL

2. Tank of gasoline: 60 L   60 mL   6 L

3. Bathtub: 300 L   30 L   3 L

4. Tube of shampoo: 50 mL.   500 mL   5 L

5. Can of cola: 36 L   36 mL   360 mL

6. Large pitcher: 20 mL   200 mL   2000 mL

7. Bottle of perfume: 3 mL   30 mL   300 mL

8. Kitchen sink: 4 L   40 L   400 mL

Convert each measure.

9. 2.9 L = _____ mL

10. 600 mL = _____ L

11. 1200 mL = _____ L

12. 9 L = _____ mL

13. .06 L = _____ mL

14. 45 mL = _____ L

15. 8 mL = _____ L

16. .058 L = _____ mL

17. 6.3 L = _____ mL

18. 250 mL = _____ L

19. 2000 mL = _____ L

20. 14 L = _____ mL

21. .29 L = _____ mL

22. 67 mL = _____ L

CONVERSIONS OF CUSTOMARY CAPACITY

REMEMBER: A fluid ounce is different from the ounce that measures weight. Multiply when converting to a smaller unit. Divide when converting to a larger unit. Use the table below.

1 gallon = 4 quarts   1 pint = 2 cups
1 gallon = 8 pints   1 pint = 16 fluid ounces
1 quart = 2 pints   1 cup = 8 fluid ounces
1 quart = 4 cups

EXAMPLES **6 gallons = ___48___ pints**   **(6 x 8 = 48)**

16 cups = ___4___ quarts   **(16 ÷ 4 = 4)**

Convert by dividing or multiplying.

1. 9 gallons = _____ quarts

2. 18 cups = _____ pints

3. 48 fluid ounces = _____ pints

4. 9 gallons = _____ pints

5. 56 fluid ounces = _____ cups

6. 13 quarts = _____ pints

7. 9 quarts = _____ cups

8. 28 quarts = _____ gallons

9. 64 pt = _____ gal

10. 76 pt = _____ cups

11. 13 pt = _____ fl oz

12. 22 pt = _____ qt

13. 92 cups = _____ qt

14. 17 cups = _____ fl oz

15. 19 gal = _____ qt

16. 58 cups = _____ pt

17. 72 fl oz = _____ pt

18. $5\frac{1}{2}$ gal = _____ pt

19. 34 fl oz = _____ cups

20. $2\frac{1}{2}$ qt = _____ pt

21. $7\frac{1}{2}$ qt = _____ cups

22. 37 qt = _____ gal

23. 44 pt = _____ gal

24. $3\frac{1}{2}$ pt = _____ cups

25. $2\frac{1}{4}$ pt = _____ fl oz

26. 23 pt = _____ qt

27. 5 gal = _____ cups

28. 128 fl oz = _____ qt

Conversion of Time Units

Multiply when converting to a smaller unit. Divide when converting to a larger unit. Use the table below.

1 year = 12 months	1 day = 24 hours	1 week = 7 days
1 year = 365 days	1 hour = 60 minutes	
1 month = 30 days	1 minute = 60 seconds	

Note: Some months have 31, 28, or 29 days. 30 days is used as an approximation.

EXAMPLES

6 years = 72 months

  12
x 6
72 months

4 hours = 14,400 seconds

  60
x 4
240 minutes

  240
x 60
14,400 seconds

Convert by dividing or multiplying. (≈ means "approximately equal to".)

1. 3 years = _____ days

2. 120 days ≈ _____ months

3. 8 weeks = _____ days

4. 144 hours = _____ days

5. 12 hours = _____ minutes

6. 300 seconds = _____ minutes

7. 8 years = _____ months

8. 730 days = _____ years

9. 7 months ≈ _____ days

10. 49 days = _____ weeks

11. 17 days = _____ hours

12. 420 minutes = _____ hours

13. 132 months = _____ years

14. 18 minutes = _____ seconds

15. 1 month ≈ _____ hours

16. 1 day = _____ minutes

17. 1 hour = _____ seconds

18. 1 week = _____ hours

19. 7200 seconds = _____ hours

20. 336 hours = _____ days

21. $2\frac{1}{2}$ days = _____ hours

22. 78 months = _____ years

23. 150 minutes = _____ hours

24. $6\frac{1}{2}$ months = _____ days

ADDING AND SUBTRACTING UNITS OF TIME

Add. Rename when necessary.

1. 9 hr 46 min
+2 hr 9 min

2. 4 min 19 sec
+3 min 29 sec

3. 1 day 19 hr
+3 days 8 hr

4. 6 hr 34 min
+1 hr 28 min

5. 8 min 53 sec
+5 min 43 sec

6. 1 day 16 hr
+1 day 8 hr

7. 9 hr 14 min
8 hr 58 min
+3 hr 41 min

8. 6 min 38 sec
2 min 57 sec
+2 min 29 sec

9. 2 days 9 hr
1 day 19 hr
+1 day 23 hr

Subtract. Borrow when necessary.

10. 8 hr 39 min
−6 hr 22 min

11. 6 min 58 sec
−2 min 9 sec

12. 16 days 4 hr
− 1 day 21 hr

13. 12 hr 24 min
− 5 hr 45 min

14. 9 min 40 sec
−6 min 50 sec

15. 4 days 2 hr
−2 days 6 hr

16. 3 hr 17 min
−2 hr 45 min

17. 8 min
− 16 sec

18. 3 days
−1 day 10 hr

19. 4 yr 3 mo
−1 yr 8 mo

20. 6 wk
−3 wk 4 days

21. 2 wk
− 1 hr

COMPUTING ELAPSED TIME

REMEMBER: There are 24 hours in a day. The morning hours (midnight to noon) are called A.M. The afternoon and evening hours (noon to midnight) are called P.M. To count the hours from morning to afternoon, you must count the time to noon and then add the afternoon time. Draw a clock, and use it if it helps. To count time from one day to later in the next day, you must add 24 hours.

EXAMPLES **Compute the number of hours and minutes between the two given times.**

From 9 A.M. to 3 P.M.

3 hr to noon
+3 hr after noon
6 hr

From 10:30 A.M. to 4 P.M.

1 hr 30 min to noon
+4 hr   after noon
5 hr 30 min

10:15 A.M. Mon. to 2:10 P.M. Tues.

24 hr
1 hr 45 min
+ 2 hr 10 min
27 hr 55 min

Compute the number of hours and minutes between the two given times.

1. 10:00 A.M. to 4:00 P.M.

2. 8:00 A.M. to 6:00 P.M.

3. 11:00 A.M. to 1:00 P.M.

4. 9:00 A.M. to 3:30 P.M.

5. 7:00 A.M. to 2:15 P.M.

6. 8:30 A.M. to 7:00 P.M.

7. 6:00 A.M.. to 5:30 P.M.

8. 10:15 A.M. to 3:15 P.M.

9. 7:15 A.M. to 2:30 P.M.

10. 9:10 A.M. to 4:40 P.M.

11. 8:15 A.M. to 4:40 P.M.

12. 11:20 A.M. to 7:45 P.M.

13. 8:30 A.M. Mon. to 11:30 A.M. Tues.

14. 7:00 A.M. Mon. to 1:00 P.M. Tues.

15. 10:50 A.M. Tues. to 2:00 P.M. Wed.

16. 7:00 A.M. Mon. to 9:00 P.M. Tues.

17. 9:30 A.M. Mon. to 11:00 P.M. Wed.

18. 12:00 noon Mon. to 1:30 P.M. Tues.

FINDING FRACTIONAL PARTS

To find a fractional part, divide the number of units in the part by the number of units in the whole. You must have the same kind of parts in the numerator and denominator; for example, dollars, or minutes, or inches.

EXAMPLE
What part of a day is 6 hours?
1 day = 24 hours, so

$$\frac{6 \text{ hours}}{24 \text{ hours}} = \frac{1}{4} \text{ day}$$

1. What part of 1 day is 12 hours?

Answer: _____

2. What part of 1 foot is 4 inches?

3. What part of 1 hour is 40 minutes?

Answer: _____

Answer: _____

4. What part of 1 week is 1 day?

5. What part of 1 year is 2 months?

Answer: _____

Answer: _____

6. What part of 1 meter is 1 centimeter?

7. What part of a minutes is 10 seconds?

Answer: _____

Answer: _____

8. What part of 1 day is 3 hours?

9. What part of 1 kilometer is 1 meter?

Answer: _____

Answer: _____

10. What part of 1 kilogram is 100 grams?

11. What part of 1 foot is 6 inches?

Answer: _____

Answer: _____

12. What part of 1 pound is 4 ounces?

13. What part of 1 year is 5 days?

Answer: _____

Answer: _____

14. What part of 2 hours is 30 minutes?

15. What part of 2 liters is 400 mL?

Answer: _____

READING A THERMOMETER

Temperatures may be odd numbers even though they are not shown on the scale.

The figures below represent partial thermometers showing the temperature at different times of day. Use the figures to answer the questions.

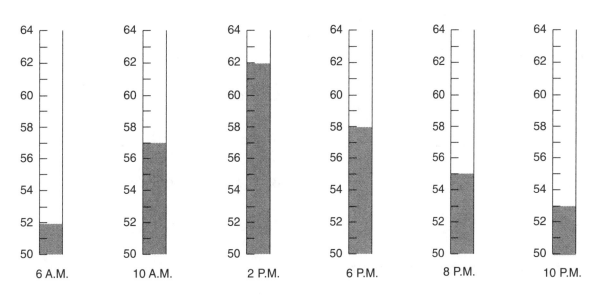

6 A.M.   10 A.M.   2 P.M.   6 P.M.   8 P.M.   10 P.M.

Write the temperature for each hour.

1. 6 A.M. _____

2. 10 A.M. _____

3. 2 P.M. _____

4. 6 P.M. _____

5. 8 P.M. _____

6. 10 P.M. _____

Write the change in temperature from one given time to another. Write + if the temperature went up and – if the temperature went down.

7. 6 A.M. to 10 A.M. _____

8. 6 A.M. to 2 P.M. _____

9. 6 A.M. to 6 P.M. _____

10. 10 A.M. to 2P.M. _____

11. 10 A.M. to 6 P.M. _____

12. 10 A.M. to 10 P.M. _____

13. 2 P.M. to 6 P.M. _____

14. 2 P.M. to 10 P.M. _____

15. 6 P.M. to 8 P.M. _____

16. 8 P.M. to 10 P.M. _____

17. At what hour was the temperature the highest? _____

18. At what hour was the temperature the lowest? _____

19. At 7 P.M., was the temperature probably about 56° or 62°? _____

USING UNITS OF TEMPERATURE

EXAMPLES **Average room temperature is about 20°C (68°F).**

Write whether the temperature is C or F.

1. A hot summer day: 35° _____

2. A cold winter day: 20° _____

3. Sick person: 102° _____

4. Arctic Ocean: 5° _____

5. Steam: 100° _____

6. Spring day: 20° _____

7. Ice cream: 30° _____

8. Swimming pool: 25° _____

9. Hot coffee: 140° _____

10. Iced tea: 40° _____

Circle the most sensible temperature.

11. Orange juice: 45°C   10°C   125°C

12. Hot shower 80°C   100°C   30°C

13. Mountain stream: 5°C   40°C   -20°C

14. Snowball: 20°C   -5°C   5°C

15. Hot oven: 250°C   100°C   450°C

16. Ice cube: 32°C   20°C   0°C

17. Autumn day: 30°C   10°C   50°C

18. Toasted bread: 35°C   90°C   15°C

For each temperature, write the letter of the appropriate match.

19. 10,000°C _____

a. Temperature in a volcano

20. 37°C _____

b. Temperature in a cool cellar

21. 500°C _____

c. Temperature in a snowstorm

22. 12°C _____

d. Temperature on the sun

23. 45°C _____

e. Recommended room temperature

24. -5°C _____

f. Temperature on a hot desert

25. 20°C _____

g. Normal body temperature

CHAPTER 5

POSTTEST

Estimate the length of the line below in inches and also in centimeters.

_____   **1.** _____ in.   **2.** _____ cm

3. The length of a track for horse races is about (check one)

2 meters _____   2 kilometers _____   2 millimeters _____

Convert each of the following lengths to meters by multiplying or dividing.

4. 7.2 kilometers = _____ meters   **5.** 42 cm = _____ m

Convert each of the following to feet.

6. 9 yd = _____ ft   **7.** 96 in. = _____ ft   **8.** 5 mi = _____ ft

Perform the operations indicated.

9. 2' 5"   **10.** 6' 9"   **11.** 7' 2"   **12.** 3' 8"
     +9' 6"   +1' 8"   −5' 7"   x   4

13. Find the perimeter of the square at the right. Write the answer in feet and inches.

5"

Perimeter _____

14. A raisin weighs about (check one)   1 gram _____   1 kilogram _____

Convert each of the following.

15. 9.1 kg = _____ g   **16.** 216 grams = _____ kilograms

17. 16 lb = _____ oz   **18.** 192 oz = _____ lb   **19.** 7.8 tons = _____ lb

20. A bucket might hold (check one)   8 mL _____   8L _____

Convert each measure.

21. 17 gal = _____ qt

22. 52 pints = _____ quarts

Convert each time.

23. 480 sec = _____ min

24. 13 hours = _____ minutes

Perform the operations indicated.

25.   7 hr 29 min
        +6 hr 43 min

26.   4 min 48 sec
        −1 min 33 sec

27.   8 days 3 hr
        −3 days 16 hr

Find the elapsed time in each exercise.

28. 9:40 A.M. to 2:50 P.M.

29. 8 A.M. Mon. to 9 A.M. Tues.

Answer: _____

Answer: _____

30. On a good day for ice skating, the air temperature might be

(check one)   25°F _____   25°C _____

31. Raymond is 75 inches tall. How tall is he in feet and inches?

Answer: _____

32. What part of an hour is 16 minutes?

Answer: _____

33. 2.5 L = _____ mL

166

PRETEST

Find the average and the median for each group of numbers.

1. The numbers: 20, 21, and 25 _____

2. Test scores: 85, 67, 92, and 76 _____

3. Find the median and mode for the following scores:
71, 74, 75, 75, 75, 81, 81, 90, 90, 91, 95. _____

INCHES OF RAINFALL IN PARK CITY

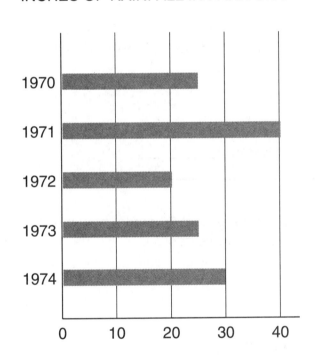

ROBERTS FAMILY BUDGET

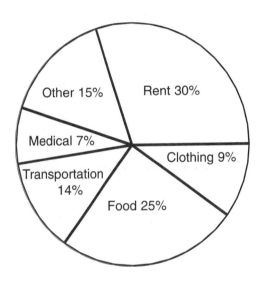

Examine the graph of rainfall in Park City during recent years.

4. In what year was the rainfall the highest? _____

5. How many inches fell during 1970? _____

6. What was the least rainfall in any one year? _____

7. What was the average rainfall for the period 1970-1974? _____

The graph shows the Roberts family budget for last year.

8. What percent did they spend on medical? _____

9. What percent was spent on food and clothing? _____

10. The family spent a total of $30,000. What did they spend on rent? _____

11. How much did they spend on transportation? _____

The table lists distances between towns A, B, C, D, and E in miles.

	A	B	C	D	E
A	—	114	231	332	465
B	114	—	165	219	326
C	231	165	—	32	124
D	332	219	32	—	89
E	465	326	124	89	—

12. How far is it from B to C? _____

13. How far is it from C to A? _____

14. At 55 miles per hour, how many hours will it take to drive from C to B?

Answer: _____

15. If a car travels 45 miles per hour for 3 hours, how far does it travel?

Answer: _____

16. If a man walks 3 miles per hour, how long will it take him to walk 20 miles?

Answer: _____

17. Find the interest on $600, at 5% for 2 years.

Answer: _____

18. If 4 apples cost 60 cents, how much does 1 apple cost?

Answer: _____

19. If there are 3 red marbles and 4 green marbles in a can, what is the probability of picking a red one?

Answer: _____

20. With one toss of a coin and one roll of a die, what is the probability of tossing a heads and rolling a 3?

Answer: _____

CALCULATING AVERAGES

To find an average, add all quantities to be averaged and then divide by the total number of quantities.

EXAMPLES
Find the average of 22, 28, and 37.

```
 22              29  Average
 28           3)87
+37
 87
```

If John received grades of 72, 79, 84, and 89 on four tests, what was his average?

```
 72              81  Average
 79           4)324
 84
+89
324
```

Find averages for the following.

1. The numbers: 15, 16, and 20

Answer: _____

2. Football scores: 7, 14, and 21

Answer: _____

3. Test scores: 88, 81, 90, and 85

Answer: _____

4. Amounts of money: $14, $31, and $15

Answer: _____

5. Distances: 100 miles, 500 miles, and 600 miles

Answer: _____

6. Times: 10 minutes, 55 minutes, 34 minutes, and 21 minutes

Answer: _____

7. Numbers of people: 1000, 2400, and 1400

Answer: _____

8. Amounts of money: $15.25, $18.75, $20.40, and $16

Answer: _____

CALCULATING AVERAGES

To find the average, or arithmetic mean, add all quantities and divide by the total number of quantities.

EXAMPLES

Find the average of 21, 29, and 23.

```
 21          24.3 ──→ 24
 29        3)73.0
+23
 73
```

Find the average of the following: $21.95, $34.56, $28.60, $25.55.

```
 $21.95         27.665 ──→ $27.66
 $34.56       4)110.660
 $28.60
+$25.55
$110.66
```

Find averages for the following. Round answers to nearest whole number or nearest cent.

1. The numbers: 18, 24, and 25

   Answer: _____

2. Distances: 59 km, 74 km, 88 km, and 61 km

   Answer: _____

3. Expenditures: $67.80, $84, and $91.88

   Answer: _____

4. Numbers of people: 5789, 3400, and 6508

   Answer: _____

5. Times: 45 min, 55 min, 75 in, and 84 min

   Answer: _____

6. Scores: 99, 76, 80, 71, and 83

   Answer: _____

7. Amounts of money: $101.50, $89.85, $95.91, and $70

   Answer: _____

8. Basketball scores: 110, 95, 114, 99, and 80

   Answer: _____

170

MEDIAN

To find the median of a set of numbers, first arrange the numbers in order from the smallest to the largest. The median is the number in the middle. If there are two numbers in the middle, the median is the average of the two.

EXAMPLES **Find the median of each set of numbers.**

5, 9, 3, 11, 7

33, 45, 29, 40, 42, 44

Place the numbers

3, 5, 7, 9, 11   in order   29, 33, 40, 42, 44, 45

7 is the middle number.

So, 7 is the median.

40 and 42 are in the middle.
41 is halfway between 40 and 42.
So, 41 is the median.

Find the median for each set of numbers.

1. 6, 9, 3, 5, 12

2. 56, 78, 60, 85, 92

3. 29, 18, 27, 24, 31, 20, 30

4. 90, 85, 92, 77, 99, 65, 75

5. 46, 40, 52, 61, 42, 48

6. 3, 5, 7, 4, 6, 8

7. 77, 88, 84, 79, 89, 84

8. $40, $38, $58, $21

9. $5.25, $6.86, $4.99, $5.99, $2.49

10. 89, 87, 88, 91, 88

11. 2, 5, 100, 7, 8

12. 1.7, 1.3, 2.1, 2.0, 1.9

13. 5.9, 4.8, 5.0, 6.1, 3.9, 4.7

14. 102, 101, 8, 104, 99, 203

MEDIAN

Find the median of each set of data.

1. Heights: 6'8", 6'10", 7'1", 6'7", 5'9"

2. Runs scored: 3, 0, 6, 1, 2, 1, 12, 5, 0, 4, 1

3. Test scores: 77, 81, 82, 75, 90, 62, 79, 83, 91, 73, 80

4. Ticket prices: $18, $5.50, $10, $75, $5.60, $21

5. Football scores: 13, 6, 6, 0, 20, 7, 0, 24

6. Points per game: 10, 7, 9, 8, 14, 2, 6

7. Prices: $49.95, $109, $88.50, $129.75, $98

8. Measurements: 8.65, 9.0, 7.9, 8.85, 8.15, 9.1, 8.5, 8.2

9. Test scores: 84, 84, 84, 90, 90, 90

10. Test scores: four 75s, three 81s. two 86s

11. Runs scored: 0 three times, 2 four times, 3 once, 5 once

12. Test scores: three 70s, four 78s, two 71s, three 75s, two 80s, three 88s, one 92

172

MODE

To find the mode of a set of numbers, list the numbers in order from the smallest to the largest. The mode is the number that appears the most number of times.

EXAMPLES **Find the mode for each set of numbers.**

2, 3, 3, 4, 5, 5, 5, 6, 7, 10

5 is the mode. It appears three times.

73, 75, 81, 81, 87, 87, 87, 87, 91, 95

87 is the mode. It appears four times.

Find the mode of each set of numbers.

1. 5, 10, 10, 12, 13

2. 70, 75, 75, 75, 81, 81, 82

3. 21, 21, 23, 34, 45

4. 65, 73, 73, 73, 84, 84, 90

5. $40, $45, $52, $52, $60, $60, $72, $60, $78

6. $76, $82, $87, $87, $88, $93, $93, $99, $99, $87

7. 2, 4, 6, 4, 5, 7, 4, 8, 9

8. 45, 50, 60, 50, 65, 75, 75, 65, 75, 85

9. 60%, 78%, 67%, 92%, 78%, 87%, 80%, 67%, 78%, 94%, 80%, 87%, 99%

10. $10, $15, $12, $10, $19, $12, $18, $19, $21, $14, $14, $15, $12

11. 80, 76, 67, 91, 45, 88, 82, 68, 96, 82, 79, 82

12. 3.5, 1.8, 2.4, 1.7, 1.8, 2.4, 1.7, 1.8

MEAN, MEDIAN, MODE

Find the median and the mode.

1. The baseball players got the following number of hits in their first three games: 0, 1, 2, 3, 4, 1, 1, 0, 0, 2, 0. Find the median and the mode.

2. Gary had the following grades on 10 point quizzes: 7, 6, 8, 7, 4, 10, 9, 7, 8, 10. Find the median and the mode.

3. Students ate the following numbers of pieces of pizza: 1, 4, 2, 3, 3, 2, 1, 3, 3. Find the mean, median, and mode.

4. Students sold the following numbers of tickets: 0, 2, 1, 3, 20, 8, 3, 2, 3, 9. Find the mean, median, and mode.

5. Basketball players scored the following number of points in their first three games: 0, 1, 6, 7, 8, 15, 15, 18, 21, 78, 71. Find the mean, median, and mode.

6. Records sold for the following prices: $5.99, $7.99, $9.99, $7.99, $6.99, $6.99, $7.99, $5.99. Find the mean, median, and the mode.

7. A week in July had the following high temperatures: 88°, 91°, 87°, 87°, 95°, 89°, 84°. Find the mean, median, and mode.

8. People were asked which of three TV programs they watched. The results were: A 267, B 78, C 123. Is the favorite a mean, median, or mode?

CALCULATING AVERAGES

If a quantity (score, amount of money, measurement) is repeated, then it must be added that many times to find an average.

EXAMPLES

If for 4 weeks you received $20, $20, $24, $27, what is the average?

$20
 20
 24
+27
$91

$$\begin{array}{r} 22.75 \\ 4\overline{)91.00} \end{array} \rightarrow \$22.75$$

Of on 10 tests you received four 70s, four 80s, one 85, and one 86, what is the average?

4 x 70 → 280
4 x 80 → 320
         85
       + 86
        771

$$\begin{array}{r} 77.1 \\ 10\overline{)771.0} \end{array} \rightarrow 77$$

Find the average, or arithmetic mean, for the following. Round to the nearest whole number or nearest dollar.

1. Test scores: four 65s and one 90

   Answer: _____

2. Salary: Five $300s, two $320s, And two $250s

   Answer: _____

3. Distances: 110 miles (twice) and 200 miles (three times)

   Answer: _____

4. Times: Four 45 minutes, three 55 minutes, and one 80 minutes

   Answer: _____

5. Numbers of people: three 550s, two 700s, and one 1000

   Answer: _____

6. Football scores: 21 (three times), 35 (twice), 10, and 3

   Answer: _____

7. Expenditures: $15 (four times), $12 (twice), and $25 (once)

   Answer: _____

8. Heights: 65 in. (ten times), 66 in. (ten times), and 67 in. (ten times)

   Answer: _____

READING AND INTERPRETING BAR GRAPHS

When a bar stops between two numbers, estimate between numbers on the scale.

BUDGETS FOR FIVE CITIES

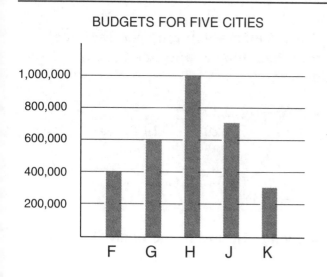

NUMBER OF POLICE OFFICERS IN NEW CITY
(Each figure represents 100.)

1. Which city has the highest budget? _____

2. What is the highest budget? _____

3. Which city has the lowest budget? _____

4. What is the difference between the highest and lowest budget? _____

5. How much higher is the budget of city H than the budget of city G? _____

6. What is the budget of city J? _____

7. What is the combined budget of cities J and K? _____

8. If the budget of city G were doubled, what would the new budget be? _____

9. If the budget of city F were cut in half, what would the new budget be? _____

10. What is the combined budget of all five cities? _____

11. Which year had the smallest number of police officers? _____

12. How many police officers were there in that year? _____

13. Which year had the largest number of police officers? _____

14. How many police officers were there in that year? _____

15. What as the increase in police from 1970 to 1980? _____

16. How many more police officers were there in 1980 than in 1950? _____

17. The police force doubled in number between the years _____ and _____ .

18. If there are the same number of teachers as police officers, how many teachers were there in 1980? _____

19. What was the total number of police officers and teachers in 1960? _____

176

READING AND INTERPRETING BAR GRAPHS

Look at the scale to find the measurement for each bar. If a bar ends between marks, estimate between numbers on the scale.

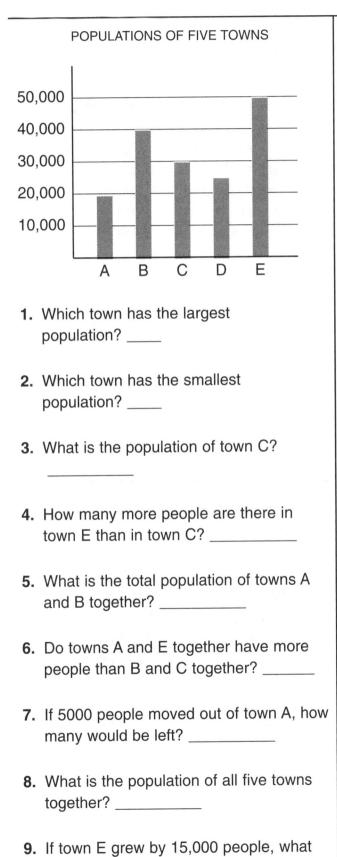

POPULATIONS OF FIVE TOWNS

BUSHELS OF CORN GROWN IN ADAMS COUNTY
(Each sack stands for 2,000 bushels.)

1. Which town has the largest population? ____

2. Which town has the smallest population? ____

3. What is the population of town C? _____

4. How many more people are there in town E than in town C? _____

5. What is the total population of towns A and B together? _____

6. Do towns A and E together have more people than B and C together? _____

7. If 5000 people moved out of town A, how many would be left? _____

8. What is the population of all five towns together? _____

9. If town E grew by 15,000 people, what would its population be? _____

10. In what year was the most corn grown? _____

11. How much was grown in that year? _____

12. In what year was the least corn grown? _____

13. How much was grown in that year? _____

14. How many bushels were grown in 1980? _____

15. How much more was grown in 1980 than in 1960? _____

16. How many bushels were grown in 1970? _____

17. Did 1950 and 1960 together produce more than 1970? _____

18. If 2005 produces twice as much as 1980, how much will be grown in 2005? _____

MAKING BAR GRAPHS

Draw two perpendicular lines for your graph. Measure and mark off distances at even intervals. Draw the bars neatly. Be sure the bars are the right length.

Draw bar graphs for the given information in each of the following.

1. The following was the annual snowfall in White Ridge for the years indicated:

1960	60 in.
1970	50 in.
1980	20 in.
1990	80 in.
2000	70 in.

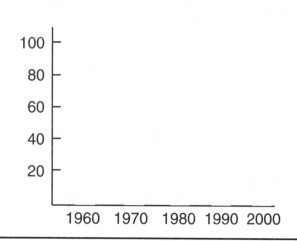

2. The following are the approximate distances from New York to the given cities:

Nashville, Tenn.	900 mi
Kansas City, Mo.	1200 mi
Denver, Colo.	1800 mi
Amarillo, Tex.	1700 mi
Reno, Nev.	2700 mi

3. The following are the approximate heights of certain mountains:

Mt. McKinley, Alas.	20,000 ft
Kirinyaga, Kenya	17,000 ft
Mt. Whitney, Calif.	14,500 ft
Mont Blanc, France	16,000 ft
Ararat, Turkey	17,000 ft

4. The following are the lengths of certain major rivers:

Rhine	800 mi
Congo	2700 mi
Mississippi	2400 mi
Colorado	1500 mi
Euphrates	1700 mi

READING AND INTERPRETING LINE GRAPHS

Read the title of the graph to find out what the information is about. Look carefully at the vertical scale to learn what each level means.

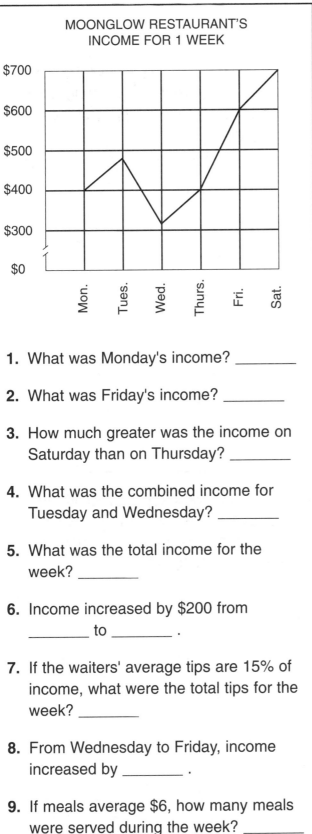

MOONGLOW RESTAURANT'S INCOME FOR 1 WEEK

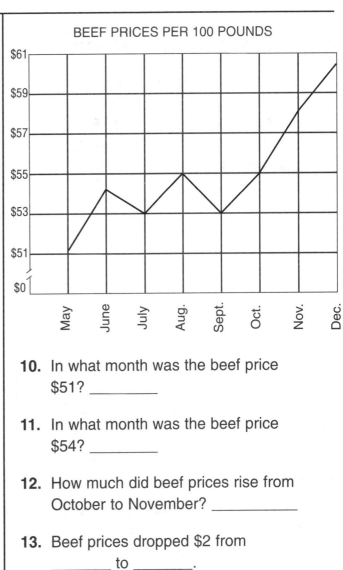

BEEF PRICES PER 100 POUNDS

1. What was Monday's income? _____

2. What was Friday's income? _____

3. How much greater was the income on Saturday than on Thursday? _____

4. What was the combined income for Tuesday and Wednesday? _____

5. What was the total income for the week? _____

6. Income increased by $200 from _____ to _____ .

7. If the waiters' average tips are 15% of income, what were the total tips for the week? _____

8. From Wednesday to Friday, income increased by _____ .

9. If meals average $6, how many meals were served during the week? _____

10. In what month was the beef price $51? _____

11. In what month was the beef price $54? _____

12. How much did beef prices rise from October to November? _____

13. Beef prices dropped $2 from _____ to _____ .

14. Beef prices rose $7 from _____ to November.

15. During the 8 months shown, beef prices rose a total of _____ .

16. The price for 200 pounds of choice beef during August was _____ .

17. Prices were the highest during _____ .

18. Prices for September were the same as prices for _____ .

MAKING LINE GRAPHS

Give your graph a title. Choose a scale that will allow you to fit everything on the graph.

Draw a line graph for each of the following.

1. These were the average temperatures for 8 months in El Centro: April 58°F, May 64°F, June 76°F, July 81°F, August 85°F, September 72°F, October 62°F, November 51°F.

2. Here is the rainfall in Willowburg during recent years: 1975, 50 in.; 1980, 74 in.; 1985, 60 in.; 1990, 68 in.; 1995, 78 in.; 2000, 65 in.

3. The census figures for Brush Prairie show the following populations: 1940, 186; 1950, 429; 1960, 480; 1970, 620; 1980, 510; 1990, 495; 2000, 525.

4. Pyric Iron Works kept this record for sales of their wood-burning stoves: 1950, 247; 1960, 329; 1970, 310; 1980, 360; 1990, 612; 2000, 980.

READING AND INTERPRETING CIRCLE GRAPHS

REMEMBER: A circle graph gives a picture of different percentages. You can get information about comparisons with a quick look. For exercises involving computation, read the question carefully and then study the graph for necessary information.

FAMILY SPENDING IN NEWTOWN

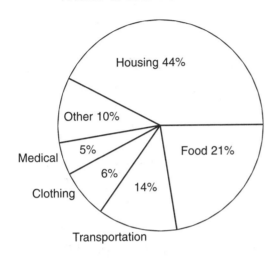

FAMILY SPENDING IN OLDTOWN

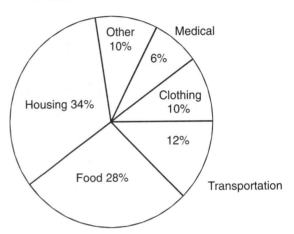

1. What is the largest expenditure? _____

2. What is the next largest expenditure? _____

3. If a family spends a total of $20,000, how much does it spend on clothing? _____

4. The amount spent on housing is about twice as much as the amount spent on _____.

5. What amount is nearly half of all spending? _____

6. If a family spends $4200 for food, what does it spend for housing? _____

7. If a family spends a total of $8000, what does it spend on transportation? _____

8. What expenditure takes 28% of the total budget? _____

9. What expenditure takes 12% of the total budget? _____

10. What two expenditures together add up to more than half of the total budget? _____ and _____

11. If the total monthly budget is $1800, what is spent on housing? _____

12. If a family has a weekly budget of $500, how much is spent on food? _____

13. If a family spends $800 per year on clothing, what is the yearly budget? _____

14. A family's monthly budget is $1000. If only 8% is spent on transportation, how much is left for other things? _____

READING AND INTERPRETING CIRCLE GRAPHS

MORGANTOWN BUDGET EXPENSES

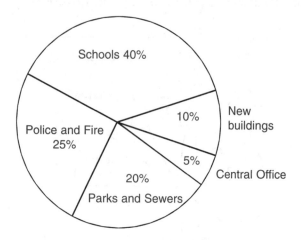

AN EMPLOYEE'S SALARY DIVISION

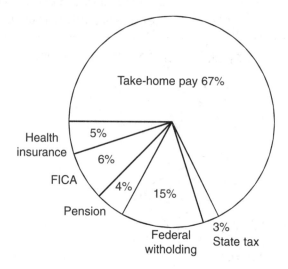

1. What is the largest expense in Morgantown's budget? _____

2. How much more of the budget is spent on schools than on parks and sewers? _____

3. If you paid $3000 in town taxes, how much of it would be spent on new buildings? _____

4. How much of every tax dollar is spent on police and fire departments? _____

5. If the total town budget is $600,000, how much is spent on each area?

Schools _____
Police and Fire Departments _____
Parks and Sewers _____
New buildings _____
Central Office _____

6. What is the total percent of all that is taken out of this person's salary? _____

7. FICA is the Social Security pension plan. What is the total percent for this and the other pension? _____

8. What is the percent taken out for federal income tax (withholding)? _____

9. What is the total percent taken out for federal and state tax? _____

10. If this person's total salary is $24,000, what is taken out for each deduction?

Federal withholding _____
FICA _____
Health insurance _____
Pension _____
State tax _____

READING AND INTERPRETING TABLES

REMEMBER: The mileage between two cities will be found by following lines across and down from those two cities. Where the lines meet, you find the distance. For example, the distance from Birmingham to Baltimore is 800 mi.

	Albany	Atlanta	Baltimore	Bangor	Birmingham	Boston	Buffalo	Charleston
Albany, NY		988	321	366	1091	170	283	712
Atlanta, GA	988		671	1315	155	1070	876	519
Baltimore, MD	321	671		632	800	400	366	391
Bangor, ME	366	1315	632		1407	233	652	1018
Birmingham, AL	1091	155	800	1407		1210	932	589
Boston, MA	170	1070	400	233	1210		458	781
Buffalo, NY	283	876	366	652	932	458		439
Charleston, WV	712	519	391	1018	589	781	439	

Find the distances between the following pairs of cities.

1. Albany and Bangor _____

2. Atlanta and Boston _____

3. Birmingham and Albany _____

4. Boston and Atlanta _____

5. Buffalo and Bangor _____

6. Charleston and Atlanta _____

7. Albany and Charleston _____

8. Boston and Birmingham _____

9. Buffalo and Atlanta _____

10. Birmingham and Charleston _____

11. Boston and Albany _____

12. Bangor and Baltimore _____

At 25 miles per gallon, how many gallons would be used for the following trips?

13. Birmingham to Baltimore _____

14. Boston to Baltimore _____

15. Birmingham to Boston _____

16. Charleston to Baltimore _____

17. Albany to Buffalo _____

18. Boston to Charleston _____

19. Buffalo to Boston _____

20. Atlanta to Birmingham _____

21. Atlanta to Boston _____

22. Bangor to Birmingham _____

USING FORMULAS

In a formula, letters take the place of numbers. When letters are next to each other, it means multiplication. If you know two out of three things in the distance formula (distance, rate, time), you can substitute to find the other.

EXAMPLES

If a car goes at a speed of 40 miles per hour (mph) for 3 hours, how far will it go?

$d = rt$ (r is rate or speed)

$d = 40 \times 3$

$d = 120$ miles

If an airplane travels a distance of 2000 miles in 5 hours, what is its average speed (rate)?

$d = rt$

$2000 = r \times 5$

$2000 \div 5 = r$

$r = 400$ miles per hour

---

1. If a car travels at a speed of 30 miles per hour, how far will it go in 5 hours?

   Answer: _____

2. If a plane travels a distance of 2500 miles in 5 hours, what is its average speed (rate)?

   Answer: _____

---

3. If a woman walks at a rate of 3 miles per hour, how far will she walk in 4 hours?

   Answer: _____

4. If a car travels 200 miles at a rate of 40 miles per hour, how long does the trip take?

   Answer: _____

---

5. If a boy rides a bike at a rate of 7 miles per hour for 3 hours, how far does he go?

   Answer: _____

6. If a girl rows a boat for 2 hours at a rate of 4 miles per hour, how far does she go?

   Answer: _____

---

7. If a bird flies south at a rate of 15 miles per hour for 40 hours, how far does it go?

   Answer: _____

8. If a 1000-mile trip takes 4 days, what is the average speed per day (rate)?

   Answer: _____

USING FORMULAS

Substitute in the formula, and use the appropriate operation to solve. Label each answer.

EXAMPLES

If a car travels 100 miles in 3 hours, what is its rate of speed?

$$d = rt$$
$$100 = 3r$$
$$100 \div 3 = r$$
$$r = 33.3 \text{ mph}$$

If a plane flies at a rate of 400 miles per hour for $3\frac{1}{2}$ hours, how far does it go?

$$d = rt$$
$$d = 400 \times 3.5$$
$$d = 1400 \text{ miles}$$

1. If a car travels 200 miles in 7 hours, what is its average rate?

   Answer: _____

2. If a woman walks at a rate of 2.4 miles per hour for 4 hours, how far does she walk?

   Answer: _____

3. If a plane travels 1125 miles at a rate of 450 miles per hour, how long does it travel?

   Answer: _____

4. If a girl rides a bike for 2 hours and 30 minutes at a rate of 8 miles per hour, how far does she go?

   Answer: _____

5. If a boat travels 45 miles at a rate of 10 miles per hour, how long does the trip take?

   Answer: _____

6. If a plane travels 1508 miles in 6 hours, what is its average rate?

   Answer: _____

7. If a butterfly migrates 2350 miles in 38 days, how far does it fly per day?

   Answer: _____

8. If a kayak is paddled 47 miles in 5 hours, what is its average rate?

   Answer: _____

USING FORMULAS

REMEMBER: The formula for interest problems is $i = prt$, in which i is interest, p is principal, r is rate, and t is time. Principal, rate, and time are multiplied to find interest.

EXAMPLES

Find the interest on $300 at 6% for 2 years.

$$i = prt$$
$$i = 300 \times .06 \times 2$$
$$i = 18 \times 2$$
$$i = \$36$$

Find the interest on $1000 at 7% for 3 years.

$$i = prt$$
$$i = 1000 \times .07 \times 3$$
$$i = 70 \times 3$$
$$i = \$210$$

Principal, rate, and time are given. Find the interest.

1. $500, 5%, 4 years.

Answer: _____

2. $800, 6%, 6 years

Answer: _____

3. $1200, 7%, 8 years

Answer: _____

4. $1500, 8%, 12 years

Answer: _____

5. $2000, 9%, 2 years

Answer: _____

6. $400, 4%, 8 years

Answer: _____

7. $3000, 5%, 10 years

Answer: _____

8. $150, 6%, 4 years

Answer: _____

9. $250, 7%, $2\frac{1}{2}$ years

Answer: _____

10. $550, 8%, $4\frac{1}{2}$ years

Answer: _____

186

WRITING RATIOS

A ratio is simply a comparison of two things by division. A ratio looks like a fraction.

EXAMPLE **Write the ratio of 6 apples to 90 cents.**

$$\frac{6 \text{ apples}}{90 \text{ cents}}$$

Simplify the ratio $\dfrac{6 \text{ apples}}{90 \text{ cents}}$.

$$\frac{\overset{1}{\cancel{6}}}{\underset{15}{\cancel{90}}} = \frac{1}{15}$$ So $\dfrac{6 \text{ apples}}{90 \text{ cents}} = \dfrac{1 \text{ apple}}{15 \text{ cents}}$

Write a ratio for the given information. Then simplify the ratio.

1. 75 cents to 5 oranges

2. 90 miles to 3 gallons

3. 27 people to 3 teams

4. $20 to 2 people

5. $7 to 3 hours

6. 600 miles to 2 hours

7. 1000 miles to 2 days

8. $0.31 to 2 cans

9. 1200 miles to 2 quarts

10. $75 to 4 people

11. $6 to 3 pounds

12. $0.75 to 3 ounces

13. Eight hundred dollars to twelve people

14. Three dollars and fifty cents to two pounds

15. Seven hundred miles to three days

16. Two hundred dollars to four hours

17. One dollar and eighty-five cents to five ounces

18. One hundred twelve miles to four gallons

UNIT RATES

If you want to find a unit rate, that goes with one of something (1 day, hour, quart, gallon, and so forth), divide both numbers in the ratio by the denominator.

EXAMPLES

If 6 apples cost 90 cents, how much will 1 apple cost

$$\frac{\overset{15}{\cancel{90}} \text{ cents}}{\underset{1}{\cancel{6}} \text{ apples}} = \frac{15 \text{ cents}}{1 \text{ apple}}$$

If a car goes 75 miles on 3 gallons, how far will it go on 1 gallon?

$$\frac{\overset{25}{\cancel{75}} \text{ miles}}{\underset{1}{\cancel{3}} \text{ gallons}} = \frac{25 \text{ miles}}{1 \text{ gallon}}$$

---

1. If 5 pencils cost 60 cents, how much does 1 pencil cost?

   Answer: _____

2. If 2 pounds of meat costs $5.80, how much does 1 pound cost?

   Answer: _____

---

3. If a car goes 80 miles in 2 hours, how far does it go in 1 hour?

   Answer: _____

4. If 2 cans of juice cost 80 cents, how much does 1 can cost?

   Answer: _____

---

5. If a car goes 162 miles on 6 gallons of gas, how far does it go on 1 gallon?

   Answer: _____

6. If a plane goes 810 miles in 3 hours, how far does it go in 1 hour?

   Answer: _____

---

7. If a man earns $43.20 in 8 hours, how much does he earn in 1 hour?

   Answer: _____

8. If 9 players spend $144 for uniforms, how much does each player spend?

   Answer: _____

---

9. If a 3-ounce can of tuna fish costs $1.02, what is the cost per ounce?

   Answer: _____

10. If a car goes 348 miles on 12 gallons of gas, how many miles does it get per gallon?

    Answer: _____

188

SOLVING PROPORTIONS

A proportion is an equality of two ratios. Be sure each fraction or ratio is set up in the same way; for example,

$$\frac{mi}{gal} = \frac{mi}{gal}$$ Solve by cross multiplying.

EXAMPLE

If a car goes 180 miles in 6 hours, how far will it go in 10 hours?

$$\frac{6 \text{ hr}}{180 \text{ mi}} = \frac{10 \text{ hr}}{n \text{ mi}}$$

6 x n = 10 x 180
6 x n = 1800
n = 300 mi

---

1. If a plane goes 900 miles in 3 hours, how far will it go in 8 hours?

Answer: _____

2. If a car goes 150 miles on 5 gallons of gas, how far will it go on 7 gallons?

Answer: _____

---

3. If 3 vans of soup cost $1.40, how much will 9 cans cost?

Answer: _____

4. If 5 gallons of gas cost $7.50, how much will 7 gallons cost?

Answer: _____

---

5. If 4 ounces of cheese cost $2.48, how much will 9 ounces cost?

Answer: _____

6. If a man earns $35.20 in 8 hours, how much will he earn in 5 hours?

Answer: _____

---

7. If a hiker walks 56 miles in 6 days, how far will she walk in 9 days?

Answer: _____

8. If it takes 48 ounces of copper for 9 couplings, how many ounces of copper are needed for 12 couplings?

Answer: _____

SOLVING PROPORTIONS

Set up each ratio in the same way:

$$\frac{mi}{gal} = \frac{mi}{gal} \qquad \frac{mi}{hr} = \frac{mi}{hr} \qquad \frac{\$}{hr} = \frac{\$}{hr}$$

Solve by cross multiplying. Round answers to nearest cent or nearest tenth.

EXAMPLE

If 2 pens cost 59 cents, how much do 3 pens cost?

$$\frac{2 \text{ pens}}{59 \text{ cents}} = \frac{3 \text{ pens}}{n \text{ cents}}$$

$2 \times n = 3 \times 59$
$2 \times n = 177$
$n = 88.5$ cents
$n = 89$ cents

---

1. If 2 cans of soup cost $2.49, how much do 3 cans cost?

Answer: _____

2. If 5 gallons of gas cost $7.79, how much will 7 gallons cost?

Answer: _____

---

3. If a car goes 185 miles in 4 hours, how far will it go in 6 hours?

Answer: _____

4. If a woman earns $24.88 in 4 hours, how much will she earn in 7 hours?

Answer: _____

---

5. If 4 ounces of cheese cost $1.60, how much will 9 ounces cost?

Answer: _____

6. If a car goes 132 miles on 5 gallons of gas, how far will it go on 6 gallons?

Answer: _____

---

7. If 2 pounds of boiled ham cost $4.45, how much will 5 pounds cost?

Answer: _____

8. If 5 acres of woods produces 8 cords of firewood, how much wood will 18 acres produce?

Answer: _____

SOLVING PROPORTIONS

REMEMBER: Set up each ratio in the same way:
$$\frac{mi}{hr} = \frac{mi}{hr}$$

EXAMPLE

If 2 pounds of fish cost $7, how much do 5 pounds cost?

$$\frac{\$7}{2\ lb} = \frac{n}{5\ lb}$$

$5 \times 7 = 2 \times n$

$35 = 2 \times n$

$n = \$17.50$

Solve by cross multiplying. Round answers to nearest cent or nearest tenth.

1. If 3 bars of soap cost $2.70, how much will 5 bars cost?

   Answer: _____

2. If a man walks 8 miles in 3 hours, how far will he walk in 5 hours?

   Answer: _____

3. If a plane goes 1700 miles in 4 hours, how long will it take to go 1000 miles?

   Answer: _____

4. If a home owner pays $35 per $1000 in taxes, what must she pay on a house worth $45,000?

   Answer: _____

5. If 3 pounds of meat cost $5.85, how many pounds can you buy for $10?

   Answer: _____

6. If a car uses 4 gallons of gas to go 90 miles, how many gallons will it use to go 150 miles?

   Answer: _____

7. If 42 acres of grazing land will support 160 sheep, how many sheep will 58 acres support? Round to the nearest whole number.

   Answer: _____

8. If 5 trees produce 23 quarts of maple syrup, how many quarts will 8 trees produce?

   Answer: _____

SOLVING PROPORTIONS

Ratios can be used to give information about large groups of populations; for example, 6 out of every 10 adults voted. Set up each ratio in the same way. Solve by cross multiplying.

EXAMPLE

If the newspaper says that 6 out of every 10 people watched a certain television show, how many watched out of 5000?

$$\frac{6}{10} = \frac{n}{5000}$$

$10 \times n = 6 \times 5000$

$10 \times n = 30,000$

$n = 3000$ people

---

1. If 8 out of every 10 students went to the dance, how many went out of 245?

   Answer: _____

2. If it rained 1 day out of every 5, how many days did it rain out of 25?

   Answer: _____

---

3. If 1 out of every 6 seniors drove, how many drove out of 300?

   Answer: _____

4. If 1 out of every 10 people is a runner, how many runners are there in 2540 people?

   Answer: _____

---

5. If 3 apples in every 8 were bad, how many were bad in 200?

   Answer: _____

6. If 7 out of every 10 people were not watching a certain television show, how many out of 2000 were not watching?

   Answer: _____

---

7. If 3 out of every 5 days were clear, how many clear days were there in a year?

   Answer: _____

8. If 3 out of every 4 apples are red, how many apples out of 128 are red?

   Answer: _____

PROBABILITY

For something that happens only once, the probability of any particular result is the fraction showing all favorable results out of all possible ones.

EXAMPLE **Find the probability of selecting a 2 or a 3.**

$P(2 \text{ or } 3) = \frac{8}{12} \text{ or } \frac{2}{3}$

The probability is $\frac{2}{3}$.

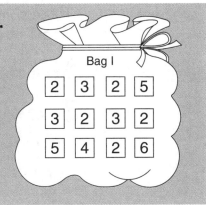

Bag I

2	3	2	5
3	2	3	2
5	4	2	6

Find the probability of selecting each set from Bag I. P(not 2) is the probability of selecting a number that is not a 2.

1. P(2)   **2.** P(3)   **3.** P(4)   **4.** P(5)

5. P(6)   **6.** P(7)   **7.** P(3 or 4)   **8.** P(5 or 6)

9. P(2 or 6)   **10.** P(1 or 2)   **11.** P(3 or 5)   **12.** P(3, 5, or 6)

13. P(not 2)   **14.** P(not 3)   **15.** P(even number)   **16.** P(odd number)

Find the probability of each outcome from the spinner.

17. P(red)   **18.** P(green)   **19.** P(white)

20. P(red or blue)   **21.** P(red or green)   **22.** P(not red)

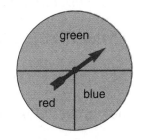

green

red | blue

Find the probability of selecting each set from Bag II.

23. P(E)   **24.** P(C)   **25.** P(H)

26. P(A or C)   **27.** P(H or R)   **28.** P(not E)

29. P(any letter)   **30.** P(vowel)   **31.** P(consonant)

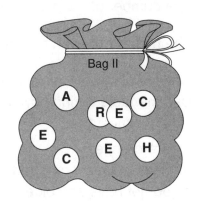

Bag II

A   R E   C
E
C   E   H

PREDICTIONS

REMEMBER: For something that happens more than once, you can predict the number of favorable outcomes. Multiply the probability of the event by the number of repetitions.

EXAMPLE **If the spinner is spun 40 times, about how many reds should appear?**

$\frac{3}{8}$ of the spinner is red, so

$\frac{3}{8}$ x 40 = 15

About 15 reds should appear.

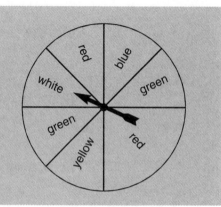

Use the picture of the spinner to predict.

1. The number of yellows in 40 spins.

2. The number of greens in 40 spins.

3. The number of pinks in 40 spins.

5. The number of blues in 100 spins.

6. The number of reds in 100 spins.

7. The number of whites in 10 spins.

Use the die (numbered 1 - 6) to predict.

9. The number of 3's in 6 rolls.

10. The number of 5's in 6 rolls.

11. The number of odd numbers in 6 rolls.

12. The number of 4's in 30 rolls.

13. The number of 7's in 30 rolls.

14. The number of even numbers in 30 rolls.

15. Lindsey plants 30 tulips in her flower garden. Use the table to find how many tulips she should expect that
    a) are orange
    b) have at least some red
    c) do not have any blue

Color	Probability
red	0.1
orange	0.2
violet	0.2
red & blue	0.2
blue	0.3

INDEPENDENT EVENTS

REMEMBER: If two events do not affect each other, find the probability that both will happen by multiplying the probabilities of each event.

EXAMPLES **Find the probability of selecting an 8 from Bag I, then getting a heads from a coin toss.**

P(8, then heads) = P(8) x P(heads)

$$= \frac{1}{9} \times \frac{1}{2}$$

$$= \frac{1}{18}$$

Bag I

2	4	6
3	8	4
12	9	3

Use Bag I and Bag II to find the probabilities.

1. P(3, then R)

2. P(3, then R)

3. P(4, then X)

4. P(even number, then C)

5. P(8, then any letter)

6. P(6, then T)

7. P(number less than 9, then a vowel)

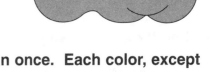

Bag II

R U R X C X U S

Find the probabilities when the spinner is spun more than once. Each color, except yellow, is one-sixth, of the circle.

8. P(blue, then green)

9. P(green, then blue)

10. P(red, then green)

11. P(red, then red)

12. P(red, then yellow)

13. P(yellow, then yellow)

14. P(red, red, red)

15. P(yellow, yellow, yellow)

16. P(red, white, blue)

17. P(green, green, green)

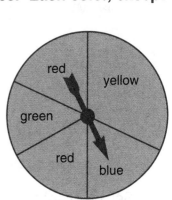

18. What is the probability of tossing a coin twice and having heads result both times?

DEPENDENT EVENTS

If two events do affect each other, use the result of the first event to help find the probability of the second event.

EXAMPLE

Find the probability of selecting a 4 and then a 3 from Bag I (without putting the 4 back in the bag).

$$P(4, \text{ then } 3) = \frac{2}{8} \times \frac{3}{7} = \frac{6}{56} \text{ or } \frac{3}{28}$$

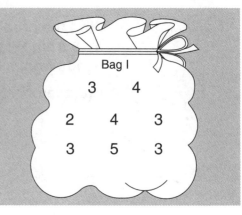

Bag I

3　　4

2　　4　　3

3　　5　　3

Find the probabilities. Assume that the first item selected is not put back into Bag I before the second item is selected.

1. P(2, then 5)　　　　2. P(5, then 2)

3. P(4, then 2)　　　　4. P(3, then 3)

5. P(4, then 4)　　　　6. P(2, then 2)

Use Bag II to find these probabilities.

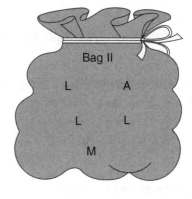

Bag II

L　　A

L　　L

M

7. P(L, then M)　　　　8. P(M, then L)

9. P(L, then L)　　　　10. P(A, then M)

11. P(A, then A)　　　　12. P(L, then L, then L)

Use Bag I and Bag II to decide which is more likely.

13. a) Selecting once and selecting an A
    b) Selecting twice and selecting two L's

14. a) Selecting once and selecting a 4
    b) Selecting twice and selecting two 4's

15. a) Selecting an odd number, then an even number
    b) Selecting an even number, then an odd number

PROBLEMS INVOLVING DIFFERENT UNITS

REMEMBER: When you need to solve a problem that has different units of measure, you can convert all the measures to the same unit.

EXAMPLE **Manny is 63$\frac{1}{2}$ in. tall. How much does he need to grow to be 6 feet tall?**

It is usually easier to write both measures with the smaller unit.

So, 6 ft = 6 x 12 = 72 inches.

72 - 63$\frac{1}{2}$ = 8$\frac{1}{2}$

Manny needs to grow 8$\frac{1}{2}$ more inches.

Table of Equivalent Measures

1 ft = 12 in.	1 lb = 16 oz	1 c = 8 fl oz	1 da = 24 hr
1 yd = 3 ft	1 ton = 2000 lb	1 pt = 2 c	1 wk = 7 da
1 mi = 5,280 ft		1 qt = 2 pt	1 yr = 12 mo
		1 gal = 4 qt	1 yr = 365 da
1 m = 100 cm	1 kg = 1000 g	1 L = 1000 mL	
1 km = 1000 m			

---

1. If 100 g of cheese cost $0.60, how much would 1 kg cost?

Answer: _____

2. Tamika buys 1 gallon of milk. After she pours four 12-fluid ounce glasses of milk, how much milk is left from the gallon?

Answer: _____

3. If a baby is 8 weeks old, how long is it until the baby's first birthday?

Answer: _____

4. LeRon weighed 8 pounds when he was born. Stanley weighed 6 lb 10 oz. How much more did LeRon weigh?

Answer: _____

5. Three shelves are cut from a board that is 10 feet long. Each shelf is 28 inches long. How much of the board is left?

Answer: _____

6. Nikki just had her twenty-fifth birthday. Is she 10,000 days old yet?

Answer: _____

PROBLEMS INVOLVING DIFFERENT UNITS

Table of Equivalent Measures

1 ft = 12 in.	1 lb = 16 oz	1 c = 8 fl oz	1 da = 24 hr
1 yd = 3 ft	1 ton = 2000 lb	1 pt = 2 c	1 wk = 7 da
1 mi = 5,280 ft		1 qt = 2 pt	1 yr = 12 mo
		1 gal = 4 qt	1 yr = 365 da
1 m = 100 cm	1 kg = 1000 g	1 L = 1000 mL	1 yr = 52 wk
1 km = 1000 m			

---

7. Harold can walk about 100 m in one minute. At that rate, how long would it take him to finish a 10-kilometer walk-a-thon?

Answer: _____

8. If a light truck holds a load of $2\frac{1}{2}$ tons, how many pounds will it hold?

Answer: _____

---

9. Mr. Matsuda is putting wallpaper in his bathroom. If each sheet is 42 inches wide, how many sheets are needed to cover a wall that is 7 feet long?

Answer: _____

10. Presidential elections are held every four years. How many weeks are there between elections?

Answer: _____

---

11. How many 1-cup glasses of lemonade can be poured from a pitcher holding two quarts?

Answer: _____

12. The rows in Mr. Clark's garden are 20 yards long. He is planting one row of tomatoes. The tomato plants are 18 inches apart. How many tomato plants does he need?

Answer: _____

POSTTEST

Find the average and the median for each group of numbers. Round to the nearest whole number if necessary.

1. The numbers: 28, 23, and 24 _____

2. Test scores: 67, 71, 93, and 88 _____

3. Find the median and mode for the following scores:
   76, 77, 81, 81, 84, 84, 84, 90, 91, 91, 92.

AVERAGE MONTHLY
TEMPERATURES IN MARKHAM

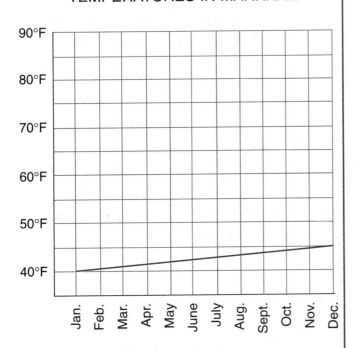

RIVER CITY BUDGET

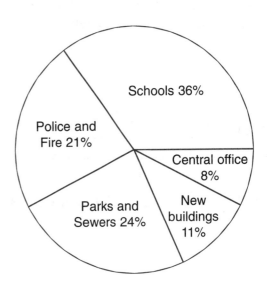

4. Which month in Markham was the hottest? _____

5. What was the average temperature in April? _____

6. What was the coolest average temperature? _____

7. What was the average temperature for the entire year? _____

8. What percent was spent on parks and sewers? _____

9. What percent went to new buildings and schools? _____

10. The city spent $20,000,000. How much was spent for police and fire? _____

11. How much of $20,000,000 was spent for new buildings? _____

The table lists distances between towns W, X, Y, and Z in miles.

	W	X	Y	Z
W	—	128	289	315
X	128	—	220	147
Y	289	220	—	193
Z	315	147	193	—

12. How far is it from W to Z? _____

13. How far is it from Z to X? _____

14. At 55 miles per hour, how many hours will it take to drive from X to Y?

Answer: _____

---

15. If a car travels 35 miles per hour for 5 hours, how far does it travel?

Answer: _____

16. If a man walks 4 miles per hour, how long will it take him to walk 23 miles?

Answer: _____

---

17. Find the interest on $800, at 4% for 3 years.

Answer: _____

18. If 5 apples cost 85 cents, how much does 1 apple cost?

Answer: _____

---

19. If there are 3 apples and 2 oranges in a bag, what is the probability of picking an apple?

Answer: _____

20. If you toss a coin twice, what is the probability of 2 heads?

Answer: _____

PRETEST

In the following exercises, write whether the figure is a square, a circle, a triangle, or a rectangle. Then find the area of each figure.

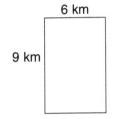

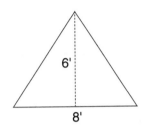

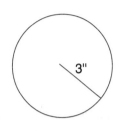

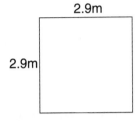

1. _____

2. _____

3. _____

4. _____

5. Area = _____

6. Area = _____

7. Area = _____

8. Area = _____

A certain circle has a radius of 1 inch.

9. Find the diameter of the circle. _____

10. Find the perimeter (circumference). _____

11. Mr. Rubin bought carpet for a room 20 feet by 30 feet.
The carpet cost $0.89 per square foot.
What was the total cost of the carpet?

Answer: _____

12. Find the volume of a box with height 6in., width 3in., and length 4.6 in.

Answer: _____

13. Train tracks are usually (check one) vertical _____ horizontal _____ .

14. They usually run (check one) parallel _____ perpendicular _____ .

Write whether each angle is acute, obtuse, or right.

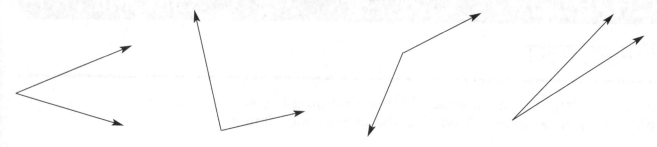

15. _____ **16.** _____ **17.** _____ **18.** _____

Tell whether each figure is *congruent* **(same size and shape) or** *similar* **(same shape) to triangle** *ABC* **below.**

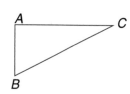

19. _____   **20.** _____

Find the square root.

21. $\sqrt{64}$ = _____   **22.** $\sqrt{\frac{1}{9}}$ = _____

Find the number of degrees in the angle marked *x.*

23.

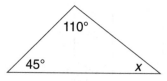

24.

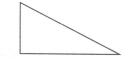

25. Use the Pythagorean rule to find the length of the third side of the triangle.

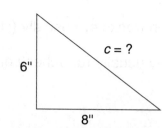

c = _____

RECOGNIZING PLANE FIGURES

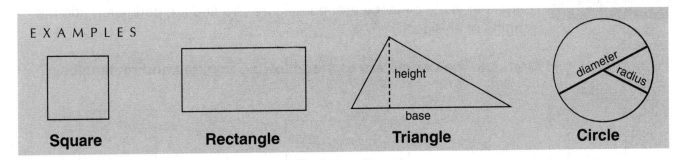

EXAMPLES

Square **Rectangle** **Triangle** **Circle**

Identify each figure. Write square, rectangle, triangle, or circle.

1. _____

2. _____

3. _____

4. _____

5. _____

6. _____

7. _____

8. _____

For each triangle, give the height and the base.

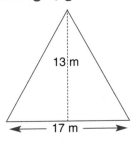

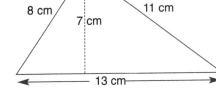

9. Height = _____   **10.** Base = _____   **11.** Height = _____   **12.** Base = _____

For each circle, give the radius and the diameter.

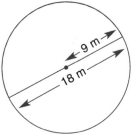

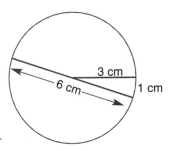

13. Radius = _____

14. Diameter = _____

15. Radius = _____

16. Diameter = _____

PERIMETERS OF SQUARES AND RECTANGLES

REMEMBER: Perimeter is the distance around a figure. To find the perimeter, add the lengths of all sides.

EXAMPLES **Compute the perimeters of the following squares and rectangles.**

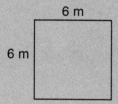
6 m
6 m

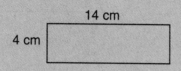
14 cm
4 cm

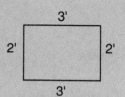
3'
2'   2'
3'

6 + 6 + 6 + 6 = 24 m   **14 + 4 + 14 + 4 = 36 cm**   **2' + 3' + 2' + 3' = 10'**

Compute the perimeter of the following squares and rectangles.

1.
5"
5"

2.
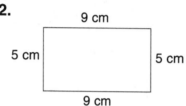
9 cm
5 cm   5 cm
9 cm

3.
39 mm
39 mm

4.
1.8 cm
.7 cm

5.
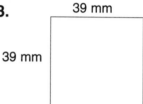
105 m
51 m

6.
$\frac{1}{8}$"
$\frac{1}{8}$"

7.
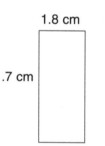
.55 m
.6 m

8.
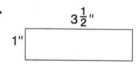
$3\frac{1}{2}$"
1"

9.

1' 2"
2' 2"

10.
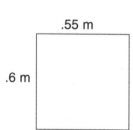
2.3 cm
2.3 cm

11.
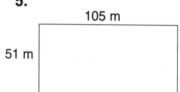
5.6 cm
2.7 cm
2.9 cm   5.5 cm
2.8 cm
2.7 cm

12.
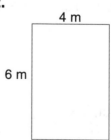
4 m
6 m

AREAS OF SQUARES AND RECTANGLES

REMEMBER: Area means measurement of the space covered by a surface, such as a floor, wall, or field. To find the area of a square or rectangle, multiply length times width.

EXAMPLES **Compute the areas of the following squares and rectangles.**

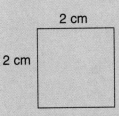

2 cm

2 cm

Area = 4 cm^2

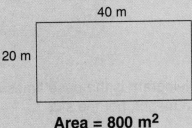

40 m

20 m

Area = 800 m^2

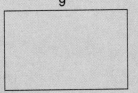

9'

6'

Area = 54 ft^2

Compute the areas for the following squares and rectangles.

1.

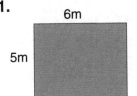

6m

5m

Area = _____

2.

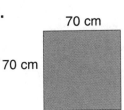

70 cm

70 cm

Area = _____

3.

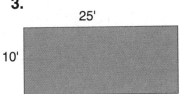

25'

10'

Area = _____

4.

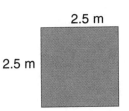

2.5 m

2.5 m

Area = _____

5.

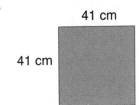

41 cm

41 cm

Area = _____

6.

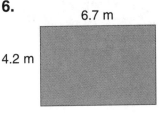

6.7 m

4.2 m

Area = _____

7.

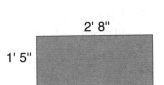

2' 8"

1' 5"

Area = _____

8.

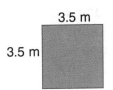

3.5 m

3.5 m

Area = _____

9.

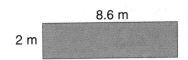

8.6 m

2 m

Area = _____

AREA OF A PARALLELOGRAM

REMEMBER: A parallelogram can be changed into a rectangle.

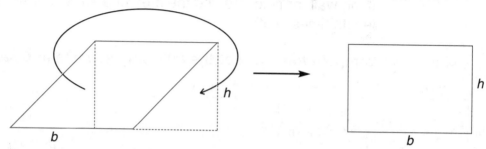

The area of a parallelogram is the base times the height.
Be sure you use the height, not the side.

EXAMPLES **Find the areas of the following parallelograms.**

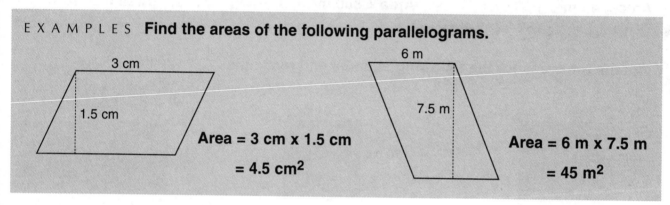

Area = 3 cm x 1.5 cm

= 4.5 cm²

Area = 6 m x 7.5 m

= 45 m²

Find the area of each parallelogram.

1.

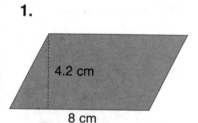

4.2 cm

8 cm

Area = _____

2.

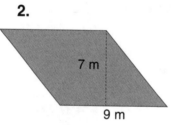

7 m

9 m

Area = _____

3.

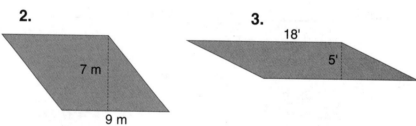

18'

5'

Area = _____

4. Base = 10m
Height = 8.6 m
Area = _____

5. Base = 73"
Height = 15"
Area = _____

6. Base = 56 m
Height = 6.1 m
Area = _____

7. Base = 9.3 m
Height = .01 m
Area = _____

8. Base = 3"
Height = 6"
Area = _____

9. Base = .3 m
Height = 1.15 m
Area = _____

10. Base = $5\frac{1}{2}$'
Height = $3\frac{1}{2}$'
Area = _____

11. Base = 14.1 cm
Height = 3.2 cm
Area = _____

12. Base = 2'6"
Height = 9"
Area = _____

206

AREA OF A TRIANGLE

REMEMBER: The area of a triangle equals one-half the base times the height.

$$A = \frac{1}{2} bh$$

EXAMPLES **Find the areas of the following triangles.**

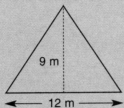

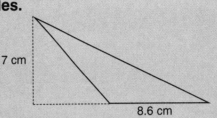

Area $= \frac{1}{2}$ **x (12 m x 9 m)**

$= \frac{1}{2}$ **x (108 m^2)**

$= $ **54 m^2**

$= .5$ **x (8.6 cm x 7 cm)**

$= .5$ **x (60.2 cm^2)**

$= $ **30.1 cm^2**

Find the areas of the following triangles.

1.

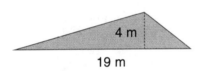

Area = _____

2.

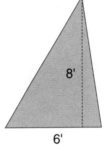

Area = _____

3.

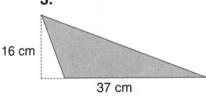

Area = _____

4. Base = 70 km
Height = 29 km
Area = _____

5. Base = 5.5 m
Height = 8 m
Area = _____

6. Base = 54'
Height = 6'
Area = _____

7. Base = 2.5 cm
Height = 15 cm
Area = _____

8. Base = 2'
Height = 1$\frac{1}{2}$'
Area = _____

9. Base = 3.2 cm
Height = 0.2 cm
Area = _____

10. Base = 6.7 m
Height = 3.9 m
Area = _____

11. Base = 6'
Height = 4'
Area = _____

12. Base = 70 km
Height = 10 km
Area = _____

13.

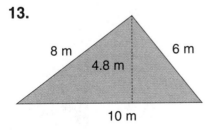

Area = _____

14.

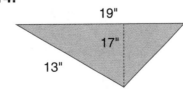

Area = _____

15.

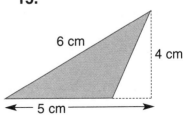

Area = _____

AREA OF A RIGHT TRIANGLE

REMEMBER: The area of a triangle is one-half the base times the height. In a right triangle, the base and the height are the sides that form the right angle.

EXAMPLES **Find the area of each of the following.**

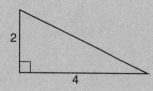

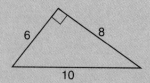

$$\text{Area} = \tfrac{1}{2}\ bh$$
$$= \tfrac{1}{2} \times (4 \times 2)$$
$$= \tfrac{1}{2} \times 8$$
$$= 4$$

$$\text{Area} = \tfrac{1}{2} \times (6 \times 8)$$
$$= \tfrac{1}{2} \times (48)$$
$$= 24$$

Find the areas of each of the following.

1.

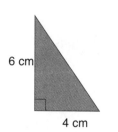

6 cm

4 cm

Area = _____

2.

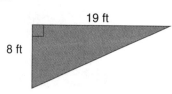

19 ft

8 ft

Area = _____

3.

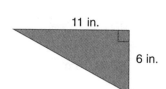

11 in.

6 in.

Area = _____

4.

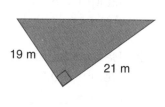

19 m

21 m

Area = _____

5.

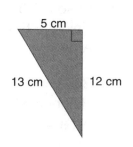

5 cm

13 cm   12 cm

Area = _____

6.

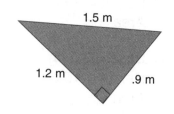

1.5 m

1.2 m

.9 m

Area = _____

For each triangle, *a* and *b* are the base and height. Find the area.

7. a = 5 cm, b = 8 cm

8. a = 9 m, b = 14 m

9. a = 76 in., b = 51 in.

10. a = 13 in., b = 19 in.

11. a = 3 cm, b = 4.6 cm

12. a = 1.0, b = 3.1

13. a = 3 m, b = 4 m

14. a = 10 m, b = 24 m

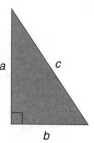

a   c

b

208

RADIUS AND DIAMETER

REMEMBER: The radius of a circle is the distance from the center to the outside.
The diameter is the distance from one side to the other, through the center.
The diameter of a circle is twice as large as the radius.

EXAMPLES

What is the radius? The diameter is 10 centimeters.

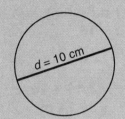

$r = \dfrac{10}{2}$

$r = 5$ cm

What is the diameter? The radius is 17 inches.

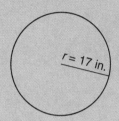

$d = 2 \times 17$

$d = 34$ in.

Give the diameter and the radius of each circle.

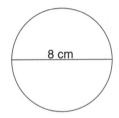

1.

Radius = _____

Diameter = _____

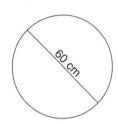

2.

Radius = _____

Diameter = _____

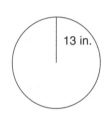

3.

Radius = _____

Diameter = _____

Find the diameter.

4. Radius = 16 cm
$d =$ _____

5. Radius = 17 m
$d =$ _____

6. Radius = 32 m
$d =$ _____

7. Radius = 176 ft
$d =$ _____

8. Radius = 89 in.
$d =$ _____

9. Radius = 300.5 cm
$d =$ _____

10. Radius = 1.7 in.
$d =$ _____

11. Radius = 18.2 cm
$d =$ _____

12. Radius = 8 cm
$d =$ _____

Find the radius.

13. Diameter = 44 cm
$r =$ _____

14. Diameter = 60 m
$r =$ _____

15. Diameter = 28 in.
$r =$ _____

16. Diameter = 51 m
$r =$ _____

17. Diameter = 17 m
$r =$ _____

18. Diameter = 20.6 cm
$r =$ _____

CIRCUMFERENCE OF A CIRCLE

The circumference is the perimeter of, or distance around, a circle. The number π is about equal to 3.14. To find the circumference, multiply π times the diameter: $C = \pi \times d$.

EXAMPLES

The diameter is 4 centimeters. Find the circumference.

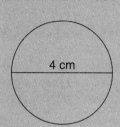

$C = \pi \times d$
$= 3.14 \times 4$
$= 12.56$

The radius is 8 meters. Find the circumference.

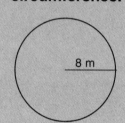

If $r = 8$, then
$d = 16$.
$C = \pi \times d$
$= 3.14 \times 16$
$= 50.24$ m

Find the circumference of each circle.

1.

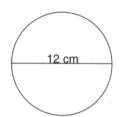

Circumference = _____

2.

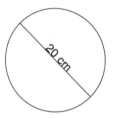

Circumference = _____

3.

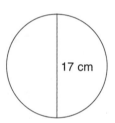

Circumference = _____

4.

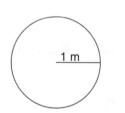

Circumference = _____

5.

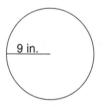

Circumference = _____

6.

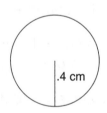

Circumference = _____

7. Diameter = 8 cm
$C =$ _____

8. Diameter = 10 in.
$C =$ _____

9. Diameter = 20 m
$C =$ _____

10. Diameter = 13 m
$C =$ _____

11. Radius = 1 ft
$C =$ _____

12. Radius = 8 cm
$C =$ _____

13. Radius = 30 m
$C =$ _____

14. Diameter = 7 in.
$C =$ _____

15. Radius = 17 ft
$C =$ _____

210

AREA OF A CIRCLE

The area is the space covered by the inside of a circle. You find the area by using 3.14 and the radius. You multiply the radius times itself and then times 3.14. We say, $A = \pi r^2$; r^2 means r x r.

EXAMPLES

Find the area of a circle whose radius is 5 centimeters.

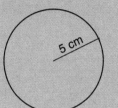

$A = \pi r^2$
$= \pi$ x r x r
$= 3.14$ x 5 x 5
$= 3.14$ x 25
$= 78.5$ cm

Find the area of a circle whose diameter is 7 meters.

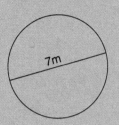

If $d = 7$, $r = 3.5$
$A = \pi r^2$
$= \pi$ x r x r
$= 3.14$ x 3.5 x 3.5
$= 3.14$ x 12.25
$= 38.465$ m^2

Find the area of each circle.

1.

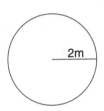

Area = _____

2.

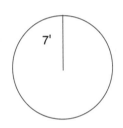

Area = _____

3.

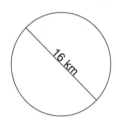

Area = _____

4. Diameter = 4 m
    Area = _____

5. Diameter = 18 m
    Area = _____

6. Radius = 1.2 km
    Area = _____

7. Radius = 9 m
    Area = _____

8. Diameter = 15"
    Area = _____

9. Diameter = 2.08 cm
    Area = _____

10. Radius = 45 m
    Area = _____

11. Diameter = 17 km
    Area = _____

12. Radius = 6'
    Area = _____

13.

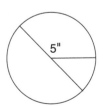

Area = _____

14.

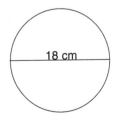

Area = _____

15.

Area = _____

PROBLEMS INVOLVING AREA

First find the area in square feet (ft²). Then multiply the area by the cost per square foot. The answer is the total cost.

EXAMPLE **The diagram is an architect's drawing of the Hoys' home. The house cost $40 per square foot. Find the area of the house and its total cost.**

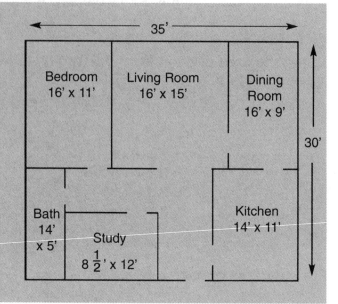

```
   35'              1050
 x 30'            x   $40
 1050 ft²         $42,000
```

1. Mrs. Hoy is buying new linoleum for the kitchen floor at $1.50 per square foot. Find the total cost of the linoleum.

   Answer: _____

2. Tile is being installed in the bathroom at $2.25 per square foot. Find the area of the bathroom and the total cost of the tile.

   Answer: _____

3. Mr. Hoy is buying paint for the dining room and kitchen ceilings. One pint of paint covers 100 square feet. How many pints will he need?

   Answer: _____

4. The living room, bedroom, and study are being carpeted for $0.78 per square foot. Find the total area and the total cost.

   Answer: _____

212

VOLUME OF A BOX-SHAPED FIGURE

REMEMBER: Volume means the amount of space inside something; for example, the amount of water that fills something up. To find the volume of a room or box, multiply length times width times height.

EXAMPLE **Find the volume.**

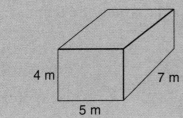

$$V = l \times w \times h$$

$$= 7 \times 5 \times 4$$

$$= 140 \text{ m}^3 \longleftarrow \text{ The symbol for cubic meters}$$

Find each volume.

1.

6 m, 6 m, 5 m

Volume = _____

2.

3 cm, 12 cm, 10 cm

Volume = _____

3.

9 in., 20 in., 13 in.

Volume = _____

4.

11 cm, 1 cm, 23 cm

Volume = _____

The length, width, and height are given. Find the volume.

5. $l = 5$ cm
$w = 2$ cm
$h = 3$ cm
$V =$ _____

6. $l = 10$ m
$w = 8$ m
$h = 7$ m
$V =$ _____

7. $l = 20$ in.
$w = 15$ in.
$h = 10$ in.
$V =$ _____

8. $l = 17$ ft
$w = 3$ ft
$h = 1$ ft
$V =$ _____

9. $l = 50$ m
$w = 1$ m
$h = 9$ m
$V =$ _____

10. $l = 1.6$ cm
$w = 3.5$ cm
$h = 2$ cm
$V =$ _____

11. $l = 5.2$ m
$w = 3$ m
$h = 4.5$ m
$V =$ _____

12. $l = 1$ ft
$w = 1$ ft
$h = \frac{1}{2}$ ft
$V =$ _____

13. A classroom is 9 meters long and 7 meters wide. The distance from the floor to the ceiling is 3 meters. Draw a figure. Find the volume.

Answer: _____

VOLUME OF A CYLINDER

REMEMBER: A cylinder is something with the shape of a can. The base, or bottom, of a cylinder is a circle. To find the volume of a cylinder, multiply the area of the base times the height. $V = \pi r^2 h$.

E X A M P L E S   **Find the volume of cylinders with the following dimensions.**

Radius = 2 cm, height = 3 cm

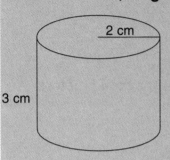

$V = \pi r^2 h$

$= (3.14)(4)(3)$

$= 37.68$ cm

Diameter = 14 m, height = 9 m

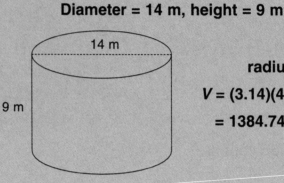

radius = 7

$V = (3.14)(49)(9)$

$= 1384.74$ m

Find the volume of each cylinder.

1. Radius = 6 m; height = 4 m

Volume: _____

2. Diameter = 16 m; height = 9 m

Volume: _____

3. Radius = 15 cm; height = 7 cm

Volume: _____

4. Radius = 6.2 cm; height = 4 cm

Volume: _____

5. Diameter = 11 cm; height = 7 cm

Volume: _____

6. Diameter = 8"; height = 10"

Volume: _____

7. Radius = 3 cm; Height = 9.5 cm

Volume: _____

8. Radius = 4 cm; height = 7 cm

Volume: _____

214

TYPES OF LINES

Parallel lines point in the same direction. Perpendicular lines form a square or right angle. Vertical lines point up and down. Horizontal lines are parallel to the ground.

EXAMPLES **A flagpole is vertical.**

A person sleeping in bed is horizontal.

A pair of train tracks is parallel.

Two streets at an intersection are usually perpendicular.

Tell whether each of the following is usually vertical, horizontal, or neither.

1. A tree trunk _____

2. A telephone wire _____

3. A tree branch _____

4. A Florida road _____

5. A table top _____

6. A skyscraper _____

7. A ray of sun _____

8. A church steeple _____

State whether each pair of items is usually parallel, perpendicular, or neither.

9. A wall and the ceiling

10. Third Avenue and Fourth Avenue

11. A flagpole and a tree trunk

_____

_____

_____

12. Football goal line and out-of-bounds line

13. Leaning Tower of Pisa and a steeple

14. A floor and a roof

_____

_____

_____

State whether each pair of arrows is parallel, perpendicular, or neither.

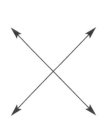

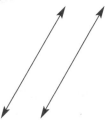

15. _____

16. _____

17. _____

IDENTIFYING TYPES OF ANGLES

REMEMBER: An angle is formed when two straight lines meet. When two lines are perpendicular, the angles formed are called *right* angles. A right angle is a 90-degree angle. An angle less than 90 degrees is called *acute*. An angle more than 90 degrees is called *obtuse*.

EXAMPLES

Draw a right angle.   **Draw an obtuse angle.**   **Draw an angle of about 45°.**

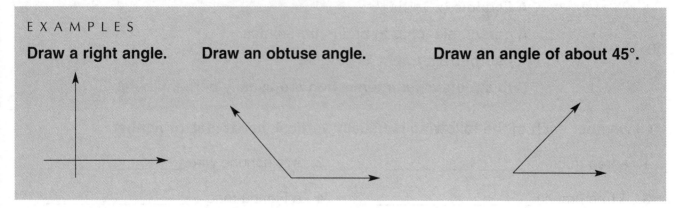

Write whether each angle is acute, right, or obtuse.

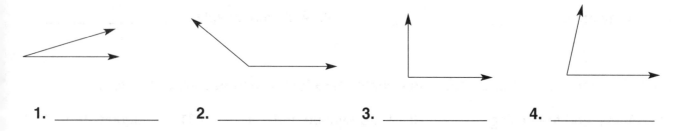

1. _____   2. _____   3. _____   4. _____

For each angle, write whether it has about 30°, 60°, 90°, or 135°.

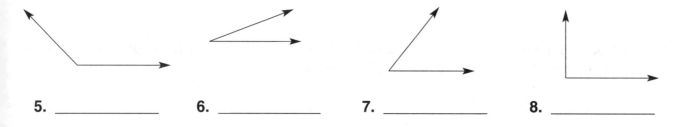

5. _____   6. _____   7. _____   8. _____

9. Draw a right angle.   **10.** Draw an acute angle.   **11.** Draw an obtuse angle.

Draw angles with about the following number of degrees.

12. 30°   **13.** 90°   **14.** 150°

216

CALCULATING ANGLE MEASURES

The measures of angles can be added, subtracted, multiplied, and divided just like any other numbers. Angles will be named by three letters on the lines forming the angle.

EXAMPLES **Using the measures of the given angles, find the number of degrees in the angle that is named.**

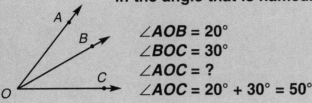

$\angle AOB = 20°$
$\angle BOC = 30°$
$\angle AOC = ?$
$\angle AOC = 20° + 30° = 50°$

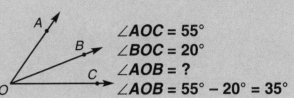

$\angle AOC = 55°$
$\angle BOC = 20°$
$\angle AOB = ?$
$\angle AOB = 55° - 20° = 35°$

Using the measures of the given angles, find the number of degrees in the other angles that are named.

1.

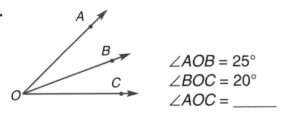

$\angle AOB = 25°$
$\angle BOC = 20°$
$\angle AOC = \underline{\hspace{1cm}}$

2.

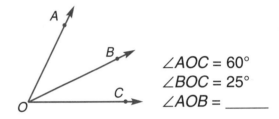

$\angle AOC = 60°$
$\angle BOC = 25°$
$\angle AOB = \underline{\hspace{1cm}}$

3.

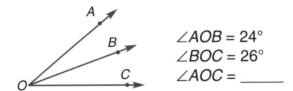

$\angle AOB = 24°$
$\angle BOC = 26°$
$\angle AOC = \underline{\hspace{1cm}}$

4.

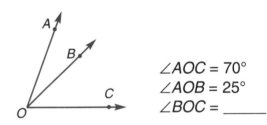

$\angle AOC = 70°$
$\angle AOB = 25°$
$\angle BOC = \underline{\hspace{1cm}}$

5.

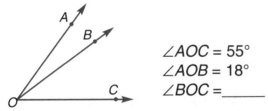

$\angle AOC = 55°$
$\angle AOB = 18°$
$\angle BOC = \underline{\hspace{1cm}}$

6.

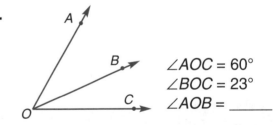

$\angle AOC = 60°$
$\angle BOC = 23°$
$\angle AOB = \underline{\hspace{1cm}}$

7.

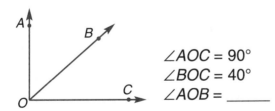

$\angle AOC = 90°$
$\angle BOC = 40°$
$\angle AOB = \underline{\hspace{1cm}}$

8.

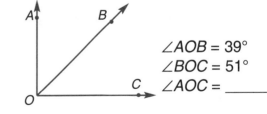

$\angle AOB = 39°$
$\angle BOC = 51°$
$\angle AOC = \underline{\hspace{1cm}}$

9.

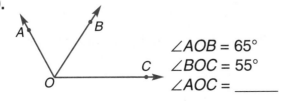

$\angle AOB = 65°$
$\angle BOC = 55°$
$\angle AOC = \underline{\hspace{1cm}}$

10.

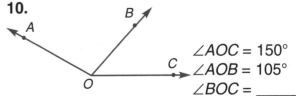

$\angle AOC = 150°$
$\angle AOB = 105°$
$\angle BOC = \underline{\hspace{1cm}}$

CONGRUENCE AND SIMILARITY

Congruent figures have exactly the same size and shape. To tell whether two figures are congruent, imagine sliding or turning one around or flipping it over to make it fit exactly on the other.

EXAMPLES   **Both pairs of figures are congruent.**

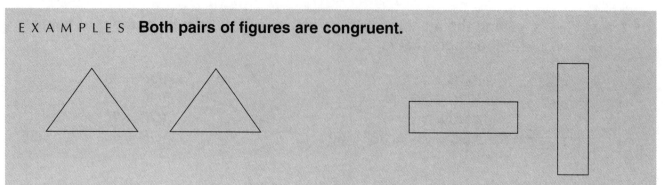

Tell whether each pair of figures is congruent. Write yes or no.

1.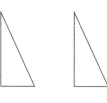

_____

2.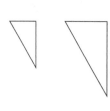

_____

3.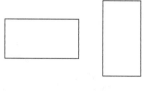

_____

4.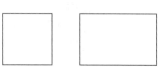

_____

5.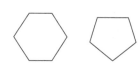

_____

6.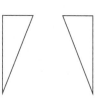

_____

Figures are *similar* if their shape is exactly the same, even if their sizes are different.

Tell whether each pair of figures is similar. Write yes or no.

7.

_____

8.

_____

9.

_____

10. Draw a figure congruent to this rectangle.

11. Draw a figure similar to this triangle.

218

ANGLES OF A TRIANGLE

A triangle has a total of 180° if you add the degrees in all three angles.

EXAMPLE **Find the number of degrees in the angle marked *x*.**

$$90 + 60 + x = 180$$
$$150 + x = 180$$
$$x = 180 - 150$$
$$x = 30 \quad \text{The angle has } 30°.$$

Find the number of degrees in each angle marked *x*.

1.

2.

3.

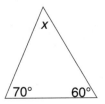

4.

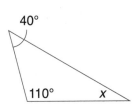

5.

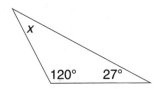

6.

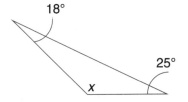

7.

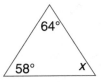

8.

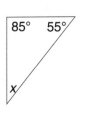

ANGLES OF A QUADRILATERAL

A quadrilateral has a total of 360° if you add the degrees in all four angles.

EXAMPLE   **Find the number of degrees in the angle marked x.**

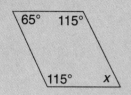

$115 + 65 + 115 + x = 360$

$295 + x = 360$

$x = 360 - 295$

$x = 65$ The angle has 65°.

Find the number of degrees in each angle marked x.

1.

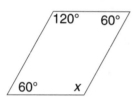

2.

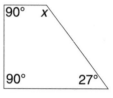

3.

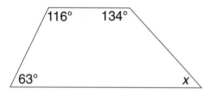

4.

5.

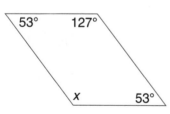

6.

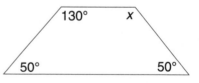

7.

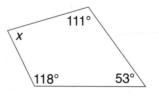

8.

220

FINDING SQUARE ROOTS

REMEMBER: The symbol $\sqrt{}$ means "the square root of." $\sqrt{9}$ is the number you multiply by itself to get 9, namely, 3. Since $3^2 = 9$. $\sqrt{9} = 3$.

EXAMPLE $\sqrt{144}$ **Try 10.**   **10 x 10 = 100**

                             Try 11.   **11 x 11 = 121**

                             Try 12.   **12 x 12 = 144** $\sqrt{144} = 12$

Find the square root.

1. $\sqrt{4} =$ _____   **2.** $\sqrt{25} =$ _____   **3.** $\sqrt{81} =$ _____   **4.** $\sqrt{100} =$ _____

5. $\sqrt{144} =$ _____   **6.** $\sqrt{64} =$ _____   **7.** $\sqrt{169} =$ _____   **8.** $\sqrt{121} =$ _____

9. $\sqrt{36} =$ _____   **10.** $\sqrt{225} =$ _____   **11.** $\sqrt{441} =$ _____   **12.** $\sqrt{196} =$ _____

13. $\sqrt{400} =$ _____   **14.** $\sqrt{529} =$ _____   **15.** $\sqrt{256} =$ _____   **16.** $\sqrt{361} =$ _____

17. $\sqrt{484} =$ _____   **18.** $\sqrt{900} =$ _____   **19.** $\sqrt{324} =$ _____   **20.** $\sqrt{289} =$ _____

21. $\sqrt{.16} =$ _____   **22.** $\sqrt{2.56} =$ _____   **23.** $\sqrt{4.84} =$ _____   **24.** $\sqrt{5.76} =$ _____

25. $\sqrt{6.76} =$ _____   **26.** $\sqrt{6.25} =$ _____   **27.** $\sqrt{\frac{1}{9}} =$ _____   **28.** $\sqrt{\frac{16}{25}} =$ _____

29. $\sqrt{\frac{9}{49}} =$ _____   **30.** $\sqrt{\frac{16}{49}} =$ _____   **31.** $\sqrt{\frac{64}{81}} =$ _____   **32.** $\sqrt{\frac{49}{121}} =$ _____

THE PYTHAGOREAN RULE

The hypotenuse is the longest side and is opposite the right angle.

EXAMPLES **In this diagram, the length of *c* can be found as follows:**

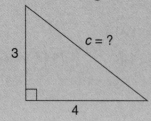

$$c^2 = 3^2 + 4^2$$
$$c^2 = 9 + 16$$
$$c^2 = 25$$
$$c = \sqrt{25}$$
$$c = 5$$

Find the value of *c* in each exercise.

1.

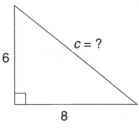

c = _____

2.

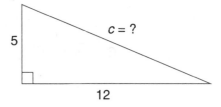

c = _____

3.

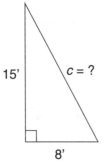

c = _____

4.

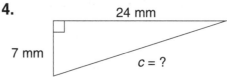

c = _____

5.

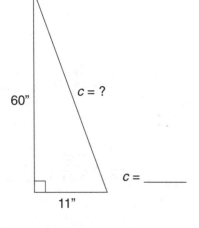

c = _____

6.

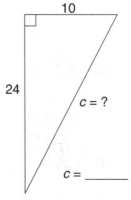

c = _____

SURFACE AREA OF A BOXED-SHAPED FIGURE

REMEMBER: The surface area of any figure is the area of all of its faces or surfaces. It may help to sketch all of them or list them.

E X A M P L E **The box below has a length of 6 in., a width of 4 in., and a height of 3 in. What is its surface area?**

Area of front and back = 2 x (6 x 3) = 36

Area of left, right sides = 2 x (4 x 3) = 24

Area of top, bottom = 2 x (6 x 4) = 48

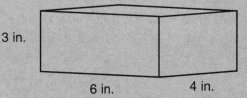

3 in.

6 in. 4 in.

36 + 24 + 48 = 108 square inches

Find each surface area.

1.

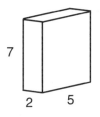

7

2   5

2.

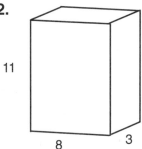

11

8   3

3.

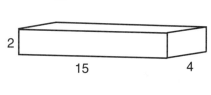

2

15   4

Surface
area = _____

Surface
area = _____

Surface
area = _____

The length, width, and height are given. Find the surface area.

4. l = 7 in.
w = 2 in.
h = 2 in.

5. l = 10 in.
w = 4 in.
h = 5 in.

6. l = 2.5 in.
w = 2 in.
h = 3 in.

7. l = 6 in.
w = 1.5 in.
h = 3 in.

S.A. = _____

S.A. = _____

S.A. = _____

S.A. = _____

SURFACE AREA OF A CYLINDER

REMEMBER: The surface area of a cylinder is the sum of the areas of the top, bottom, and curved side of the cylinder.

E X A M P L E   **Find the surface area of the cylinder with a radius of 5 in. and height of 3 in.**

Surface area = 2x (π x r^2) + 2 x π x r x h

$= 2 \times 3.14 \times 5 \times 5 + 2 \times 3.14 \times 5 \times 3$

$= 157 + 94.2$

$= 251.2$ in.^{2}

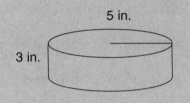

5 in.

3 in.

Find the surface area of each cylinder.

1. Radius = 6 m; height = 4 m

Surface area = _____

2. Diameter = 16 cm; height = 9 m

Surface area = _____

3. Radius = 15 cm; height = 7 cm

Surface area = _____

4. Radius = 6.2 cm; height = 4 cm

Surface area = _____

5. Diameter = 11 cm; height = 7 cm

Surface area = _____

6. Diameter = 8 in.; height = 10 in.

Surface area = _____

7. Radius = 3 cm; height = 9.5 cm

Surface area = _____

8. Radius = 4 cm; height = 7 cm

Surface area = _____

224

In the following exercises, write whether the figure is a square, a circle, a triangle, or a rectangle. Then find the area of each figure.

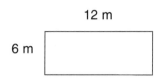

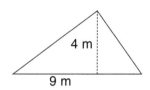

1. _____   2. _____   3. _____   4. _____

5. Area = _____   6. Area = _____   7. Area = _____   8. Area = _____

A certain circle has a radius of 3 centimeters.

9. Find the diameter of the circle. _____

10. Find the perimeter (circumference). _____

11. Donna is buying carpet for a room 25 feet by 20 feet.
    The carpet costs $0.85 per square foot.
    What is the total cost of the carpet?

    Answer: _____

12. Find the volume of a box with height 2.8', width 6', and
    length 3'.

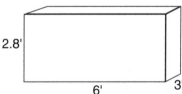

    Answer: _____

13. A tabletop is usually (check one)

    vertical _____ horizontal _____ .

14. Two roads that intersect are usually (check one)

    parallel _____ perpendicular _____ .

Write whether each angle is acute, obtuse, or right.

15. _____ **16.** _____ **17.** _____ **18.** _____

Tell whether each figure is *congruent* (same size and shape) or *similar* (same shape) to triangle *ABC* below.

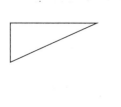 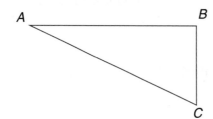 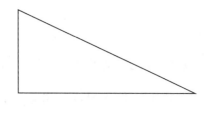

19. _____ **20.** _____

Find the square root.

21. $\sqrt{36}$ = _____

22. $\sqrt{\dfrac{1}{25}}$ = _____

Find the number of degrees in each angle marked *x*.

23.

24.

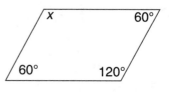

25. Use the Pythagorean Rule to find the length of the third side of the triangle.

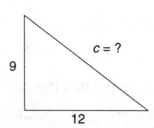

c = _____

PRETEST

1. On the number line which number is farther from 0, ⁻3 or 2? _____

2. Use arrows on the number line to add 2 + (⁻7)

3. Add: 150 + (⁻72) _____

4. Add: (⁻12.5) + (⁻7.5) _____

5. Subtract: 19 − (⁻11) _____

6. Subtract: ⁻14 − 5 _____

Add or subtract as indicated. Show your work for each step.

7. 8 + (⁻31) + (⁻9)

_____

8. $(⁻3\frac{1}{2}) + 2\frac{1}{2} + (⁻1\frac{1}{4})$

_____

9. (⁻94) − (⁻73) − (⁻36)

_____

10. 6.2 − 11.3 − (⁻17.4) + 5.6

_____

11. Find the balance in your checking account after the following transactions: Deposit $500. Write checks for $190.25, $35.45. Deposit $75.

Balance = _____

12. Find the position of a weather balloon that starts at sea level and has the following changes: Up 2505 ft; down 68 ft; up 41 ft; down 740 ft.

Position = _____

Multiply or divide each of the following as indicated.

13. ⁻5 x 7

_____

14. ⁻42 ÷ ⁻6

_____

15. 9 x 1.5

_____

16. ⁻320 ÷ ⁻8 ÷ 5

_____

17. 15 x 9 ÷ ⁻0.3

_____

18. 1200 ÷ ⁻100 x $2\frac{1}{2}$

_____

Solve each equation.

19. $x + 5 = 19$

_____

20. $3x = ⁻15$

_____

21. $\frac{a}{5} = 2.5$

_____

22. $2x − 10 = 24$

_____

23. $3y + 15 = 3$

_____

24. $\frac{z}{3} − 7 = ⁻2$

_____

25. Jan worked at $6.80 an hour and earned $217.60. How many hours did she work?

Answer: _____

THE NUMBER LINE

The numbers can be assigned to points on a line. The numbers are spaced according to their distance from 0.

EXAMPLES

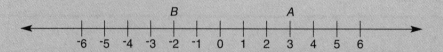

The point _A_ corresponds to the number 3. The point _B_ corresponds to number (⁻2).

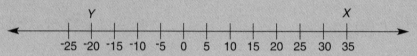

The point _X_ corresponds to the number 35. The point _Y_ corresponds to (⁻20).

Use the number line to identify the number that corresponds to each letter.

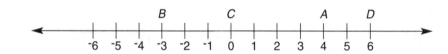

1. _A_ _____   **2.** _B_ _____   **3.** _C_ _____   **4.** _D_ _____

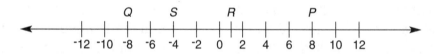

5. _P_ _____   **6.** _Q_ _____   **7.** _R_ _____   **8.** _S_ _____

On the number line below, write the following letters at the given numbers.

9. _A_ at 0   **10.** _B_ at (⁻6)   **11.** _C_ at 7   **12.** _D_ at 2.5

Write the letters at the given numbers.

13. _W_ at (⁻15)   **14.** _X_ at (35)   **15.** _Y_ at 5   **16.** _Z_ at (⁻5)

17. On the line write numbers from ⁻5 to 5.

Tell which number in each is farther from 0. If they are the same distance, write, "same."

18. 2 or 4 _____   **19.** (⁻2) or 1 _____   **20.** (⁻3) or 3 _____

228

ADDING WITH ARROWS

REMEMBER: The set of integers includes the counting numbers, their opposites, and 0. The integers can be written as points on a number line. An integer and its opposite are the same distance from 0.

EXAMPLES **You can add integers by using arrows. Arrows point in the positive or negative direction depending on the sign of the integer. The first arrow starts at 0, the second arrow starts at the endpoint of the first arrow.**

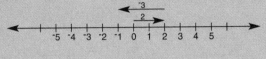

$$2 + (^-3) = (^-1)$$

$$(^-1) + (^-3) = (^-4)$$

Use arrows on the number lines to add each of the following.

1. 2 + 4

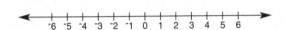

2. 6 + (⁻4)

3. (⁻2) + (⁻3)

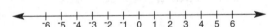

4. 2 + (⁻7)

5. (⁻3) + 3

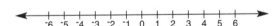

6. (⁻6) + 9

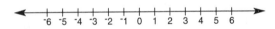

7. (⁻4) + (⁻2)

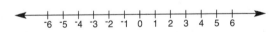

8. 3 + 2

Draw arrows to show each addition.

9. (⁻30) + 50

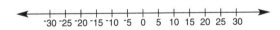

10. (⁻10) + 40

11. (⁻15) + (⁻15)

12. (⁻5) + 30

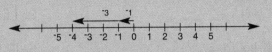

Use three arrows to add the following.

13. (⁻3) + (⁻4) + 5

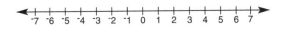

14. 6 + (⁻12) + 3

ABSOLUTE VALUE AND ADDING INTEGERS

REMEMBER: The absolute value of a number is its distance from 0 on the number line. The absolute value of 5 is 5. The absolute value of (-5) is 5. The symbol for absolute value is shown below.

EXAMPLES $|6.2| = 6.2$ $\quad\quad$ $|{-}7| = 7$

Find each absolute value.

1. $|8| = $ _____

2. $|{-}6| = $ _____

3. $|2.5| = $ _____

4. $|{-}7.2| = $ _____

5. $|92| = $ _____

6. $|{-}1| = $ _____

REMEMBER: To add two integers with the same sign, add the absolute values and use the sign of the integers. To add integers with different signs, find the difference between their absolute values and give the answer the sign of the addend with the greater absolute value.

EXAMPLES

The sum of two positive numbers is positive.	The sum of two negative numbers is negative.	The sum of integers with different signs has the sign of the one with the larger absolute value.
$3 + 15 = 18$	$(-6) + (-9) = -15$	$(-12) + 5 = {-}7$

Add the following.

7. $18 + (-7)$ _____

8. $(-56) + (-76)$ _____

9. $100 + (-78)$ _____

10. $(-5) + 5$ _____

11. $90 + 32$ _____

12. $(-45) + 55$ _____

13. $-87 + 13$ _____

14. $67 + (-32)$ _____

15. $210 + (-187)$ _____

16. $153 + 405$ _____

17. $-98 + (-45)$ _____

18. $-342 + 673$ _____

19. $76 + (-86)$ _____

20. $623 + (-236)$ _____

21. $-540 + 675$ _____

22. $84 + {-}187$ _____

23. $765 + 236$ _____

24. $-1000 + 325$ _____

For the following exercises, $A = 10$, $B = {-}25$, $C = 50$, $D = {-}20$, $E = 100$, and $F = {-}200$. Try to find each sum mentally.

25. $A + B$ _____

26. $C + D$ _____

27. $E + F$ _____

28. $A + E$ _____

29. $B + C$ _____

30. $F + A$ _____

31. $D + E$ _____

32. $C + F$ _____

33. $B + E$ _____

34. $B + B$ _____

35. $F + F$ _____

36. $E + E$ _____

37. Michael Wood has $450 in the bank. He receives a paycheck for $310 and a bill for $160. How much money will he have in the bank after he puts the paycheck in the bank and pays the bill?

Answer: _____

230

ADDING MORE THAN TWO INTEGERS

To add more than two integers, first rearrange the numbers so that the positive numbers come first. Then work from the left. Combine the positive integers two at a time. Then combine the negative integers two at a time. Then add the resulting positive and negative integers to find the final answer.

EXAMPLES
Add: 2 + (⁻8) + (⁻10)
(2) + (⁻18)
⁻16

Add: (⁻37) + 15 + (⁻12) + 9
15 + 9 + (⁻37) + (⁻12)
24   +   (⁻49)
⁻25

Add, showing your work for each step.

1. 7 + (⁻12) + 5

2. (⁻15) + (⁻10) + (⁻5)

3. 8 + (⁻33) + (⁻8)

_____

_____

_____

4. (⁻56) + 19 + 23

5. (⁻39) + 56 + (⁻14)

6. 11 + (⁻57) + (⁻4)

_____

_____

_____

Add. Do your work on another paper and show only the answer.

7. 25 + 37 + (⁻99)
_____

8. (⁻45) + (⁻32) + (⁻78)
_____

9. (⁻76) + 67 + (⁻23)
_____

10. (⁻87) + (⁻25) + 49
_____

11. 73 + 29 + 68
_____

12. (⁻36) + (⁻67) + 115
_____

13. 86 + (⁻92) + (⁻6) + 15
_____

14. (⁻90) + 37 + 52 + (⁻46) + (⁻26)
_____

15. (⁻302) + (⁻57) + 506 + (⁻12)
_____

16. 645 + 702 + (⁻371) + (⁻900)
_____

17. 75 + (⁻202) + 64 + (⁻27)
_____

18. 78 + 90 + (⁻53) + (⁻67)
_____

19. (⁻402) + (⁻78) + 47 + 620
_____

20. 56 + (⁻95) + (⁻43) + 68
_____

ADDING RATIONAL NUMBERS

REMEMBER: A rational number is a number that can be written as a ratio, or a division, of two integers. Rational numbers include:
- the integers such as 15 and -26;
- fractions and mixed numbers such as $\frac{3}{4}$ and $-\frac{7}{8}$; $6\frac{1}{2}$ and $-7\frac{3}{5}$
- decimals such as 2.25 and -5.5.

Rational numbers can be added using the same rules that are used for adding integers. If you have trouble, review the adding of decimals and fractions in Chapters 2 and 3.

EXAMPLES **Add the following.**

$6.5 + (-13.3) + 4.4$

$6.5 + 4.4 + (-13.3)$

$10.9 \quad + (-13.3)$

-2.4

$(-5\frac{1}{2}) + 8\frac{1}{4} + (-3\frac{3}{8})$

$8\frac{1}{4} + (-5\frac{1}{2}) + (-3\frac{3}{8})$

$8\frac{1}{4} + (-8\frac{7}{8})$

$-\frac{5}{8}$

Add the following rational numbers.

1. $9 + (-8.2)$

2. $-2\frac{1}{2} + (-4\frac{5}{8})$

3. $(-20) + 6.5$

4. $12.1 + (-5.5) + (-6.7)$

5. $6 + (-4\frac{1}{2}) + 2\frac{1}{4}$

6. $(-4\frac{7}{8}) + 2\frac{1}{2} + (-3\frac{1}{4})$

7. $-7.4 + 9.9 + (-11.6)$

8. $7.3 + (-3.7) + (-6.6)$

9. $2\frac{3}{4} + 2\frac{1}{2} + (-5\frac{1}{4})$

10. $(-2.25) + 5.62$

11. $8.75 + (-4.88) + (-2.04)$

12. $(-9.99) + 20 + (-1.6)$

13. $(-2\frac{1}{2}) + (-3\frac{3}{4}) + 5$

14. $6 + (-4.08) + 5.3$

15. $5 + (-6.5)$

16. $(-5\frac{1}{5}) + 2(\frac{3}{10})$

17. $4\frac{5}{8} + (-3\frac{1}{2}) + (-5)$

18. $12 + (-4.55) + (-2.66)$

19. $11\frac{1}{2} + 9\frac{3}{5} + (-1\frac{1}{2}) + 6\frac{4}{5}$

20. $67 + (-11.2) + 3.01 + (-5.6)$

21. $(-20) + 19.5 + (-8.4) + 3.7$

22. $2\frac{1}{2} + (-\frac{7}{8}) + (-3\frac{3}{4}) + 6\frac{3}{8}$

Adding Integers and Rational Numbers: Mixed Applications

Harry Johnson has a checking account with a bank. Placing money in the account is called a deposit. Writing a check to pay something takes money out of the bank. Depositing money and writing checks are called transactions. You can use positive and negative numbers to find the bank balance after each of the following sets of transactions. Each exercise starts from $0.

1. Deposit $100. Write checks for $40.25, $55.60. Deposit $75.   Balance = $79.15
2. Deposit $400. Write checks for $125.09, $46.88, $49.50.   Balance = _____
3. Deposit $650. Write checks for $90, $29.40, $188.23.   Balance = _____
4. Deposit $1000. Write checks for $250, $59. Deposit $125.   Balance = _____
5. Deposit $890. Write check for $78.95. Deposit $128.99.   Balance = _____
6. Deposit $98. Write checks for $21.08, $7.65. Deposit $34.   Balance = _____
7. Deposit $550. Write checks for $45.45, $89.76, $29.90.   Balance = _____
8. Deposit $434.55. Write check for $429.88. Deposit $99.   Balance = _____
9. Deposit $405.87. Write check for $345.22. Deposit $76.87.   Balance = _____
10. Deposit $280. Write checks for $178.25, $15.99, $21.06.   Balance = _____

Lucy Ramirez operates a deep-sea diving scientific submarine. You can calculate her position below sea level by combining integers. Sea level is 0. For "down" use negative numbers. For "up" use positive numbers. Add to find Lucy's position after each of the following dives and up or down movements. For each dive she starts at sea level.

11. Down 100 ft; up 22 ft; down 50 ft.   Position = ⁻128 ft
12. Down 88 ft; down 54 ft; up 21 ft; down 840 ft.   Position = _____
13. Down 10,405 ft; up 2900 ft; up 3040; down 509 ft.   Position = _____
14. Down 9045 ft; up 980 ft; down 780, down 678 ft.   Position = _____
15. Down 2001; up 650 ft; down 4600 ft; up 300 ft.   Position = _____

Penny and Ed Chou fly a weather balloon. Altitude is measured in feet. Find each position at the completion of the following movements.

16. Up 905 ft; down 55 ft; up 209 ft.   Position = 1059 ft
17. Up 2000 ft; down 988; down 1012 ft.   Position = _____
18. Up 3050 ft; up 256 ft; down 506 fit; down 1034 ft.   Position = _____
19. Up 900 ft; down 800 ft; down 78 ft; up 20 ft.   Position = _____
20. Up 4000 ft; down 209 ft; down 2089 ft; down 45 ft.   Position = _____

SUBTRACTING INTEGERS

REMEMBER: An addition sign (+) or a subtraction sign (–) can be the sign of an operation or the sign of a number. As the sign of an operation, + means add and – means subtract. As the sign of a number, + means a positive number and – means a negative number.

Any subtraction problem can be changed to addition by changing the sign of the operation and the sign of the number being subtracted.

The following pattern shows why it makes sense to change signs and add when subtracting a negative number:

$$10 - 4 = \quad 6$$
$$10 - 2 = \quad 8$$
$$10 - 0 = \quad 10$$
$$10 - (^-2) = \quad 12 \quad \text{Therefore } 10 - (^-2) = 10 + 2$$

EXAMPLES **Subtract by changing the sign of the operation and the sign of the number.**

$$9 - (^-4)$$
$$9 + 4 = 13$$

$$(^-7) - 8$$
$$(^-7) + (^-8) = {}^-15$$

$$(^-5) - 13 - (^-20)$$
$$(^-5) + (^-13) + 20$$
$$(^-18) \qquad + 20 = 2$$

When necessary convert the subtraction to addition.
For the first three write the related addition and then write the answer.

1. $(^-12) - 7$

2. $18 - (^-5)$

3. $(^-54) - 54$

4. $(^-60) - 17$

5. $89 - (^-21)$

6. $(^-76) - (^-12)$

7. $(^-67) - (^-67)$

8. $56 - 80$

9. $(^-32) - (23)$

Add and subtract.

10. $87 + 21 - 15$

11. $(^-65) - (^-78) + 12$

12. $101 + (^-23) - 45$

13. $(^-234) - 89 - (^-32)$

14. $(^-56) + 34 - (^-35)$

15. $(^-876) - 302 - (^-52)$

16. $(^-63) - 50 - (^-87)$

17. $(^-900) + 306 - (^-421)$

18. $(^-54) + 32 + (^-67)$

19. $89 - (^-96) - 23$

20. $(^-201) - (^-87) - (^-32)$

21. $(^-43) - 67 - 212$

22. $6 - 9 - (^-8) + (^-12)$

23. $(^-15) - 7 + 13 - (^-9)$

24. $76 + (^-75) - 15 - (^-34) + 1$

25. $(^-100) + 14 - (^-67) - 88$

234

SUBTRACTING RATIONAL NUMBERS

REMEMBER: Rational numbers follow the same rules as integers. You can change the operation and the sign of the number in order to use addition.

EXAMPLES **Add and subtract as indicated.**

$1.5 - 9.5 + 6.7$

$1.5 + (-9.5) + 6.7$

$1.5 + 6.7 + (-9.5)$

$8.2 \quad + (-9.5)$

-1.3

$(-2\frac{1}{2}) - (-1\frac{3}{4}) - 5\frac{3}{8}$

$(-2\frac{1}{2}) + 1\frac{3}{4} - 5\frac{3}{8}$

$+1\frac{3}{4} + (-2\frac{1}{2}) + (-5\frac{3}{8})$

$+1\frac{6}{8} + (-2\frac{4}{8}) + (-5\frac{3}{8})$

$+1\frac{6}{8} + (-7\frac{7}{8})$

$-6\frac{1}{8}$

Add and subtract as indicated. Rewrite using only addition for operations. Then collect positive terms first and negative terms second. Show your work for exercises 1 and 2.

1. $7.5 - 8.6 - (-10)$

2. $(-43.1) - 76.8 + 14.3 - (-12.9)$

3. $8.2 - 14.3 - (-16.4) + 3.6$

4. $-32.1 - (-78.6) - 12.9 + 56.4$

5. $90 - 143 + 63 - (-65) + (-32)$

6. $3.4 - 7.5 + (-1.7) - 6.6 - (-2.5)$

7. $2\frac{1}{2} - 4\frac{1}{2} + 3$

8. $-1\frac{1}{4} + 5\frac{3}{4} - (-2\frac{1}{4})$

9. $7\frac{1}{2} - (-3\frac{1}{4}) - 9\frac{3}{4}$

10. $-5\frac{1}{2} + 4 - 1\frac{3}{8}$

11. $9.25 - 6.05 - (-3.55)$

12. $-8\frac{5}{8} + 3\frac{1}{2} - 4\frac{3}{4}$

13. $-78.6 + (-55.4) - (-43.2)$

14. $3\frac{7}{10} - 5\frac{1}{2} + 5 - 8\frac{3}{10}$

15. $\$45.76 + \$98.02 - \$66.17$

16. $\$109.88 + (-\$34.99) - (-\$19.96)$

17. $\$178.87 - \$53.99 - (-\$7.77)$

18. $\$567.45 - \$341.05 - (\$10.90)$

19. Give an example of what subtracting amounts of money might mean in a real situation.

20. Write an example of what a negative amount of money might mean.

ADDING AND SUBTRACTING RATIONAL NUMBERS: MIXED APPLICATIONS

The following problems can be done by adding and subtracting rational numbers. A negative answer in money means someone owes money or too much money has been taken out of a bank account. A negative temperature means below 0. A negative time means before something is to happen, such as the launching of a spacecraft.

1. A bank account has a balance of $403.87. Checks are written for $209.14, $88.04, $98.20. What is the bank balance after these checks have been cashed?

    Answer: _____

2. The temperature at 9:00 A.M. is ⁻13° F. During the next three hours the temperature rises 20°. What is the temperature at 12 noon?

    Answer: _____

3. A balloon at 3,060 ft drops 490 ft and then rises 65 ft. What is the balloon's position after these movements?

    Answer: _____

4. Dwayne has $78.67 and Sherry has $178.89. They combine their money to buy a sound system that costs $209.54. How much money will they have left?

    Answer: _____

5. Lori Cushing receives a paycheck for the total amount of $356.72. The following amounts were deducted: Federal Income Tax, $31.29; Union Dues, $6.25; Credit Union, $50. What is the amount of Lori's net pay after deductions?

    Answer: _____

6. Jerry Kosky has a board 5 ft 4$\frac{1}{2}$ in. long. He cuts off 5$\frac{1}{4}$ in. from each end of the board. How long is the board after the two pieces are cut off?

    Answer: _____

7. A rocket launched from a submarine travels from 2110 ft below sea level to 14,360 ft above sea level. What is the total vertical distance traveled by the rocket?

    Answer: _____

8. Workers at the Burgher Palace work 7.5 hr with two 15 min breaks. What is the total number of hours worked?

    Answer: _____

MULTIPLICATION OF INTEGERS

A positive (+) sign can be the sign of a number or the sign of an operation. Similarly a negative (-) sign can be the sign of a number or of an operation. When you are doing multiplication, the sign always is the sign of a number. When no sign is written, the number is positive.

The product of two positive number is positive: 8 x 7 = 56

The product of a positive and a negative number is negative: 8 x (⁻7) = ⁻56

The product of two negative numbers is positive: (⁻8) x (⁻7) = 56.

It is easy to see that a positive multiplied by a negative number results in a negative number. Suppose, for example, you have $100 in the bank and you take $7 out, and you do this 8 times. 8 x (⁻7) = ⁻56 Then the total taken out is $56 and you would have to use a negative number to find your bank balance. Your final balance would be: $100 + (⁻$56) = $44.

EXAMPLES

The best way to see that the product of two negative numbers should be positive is to consider patterns.

⁻5 x 2 = -10
⁻5 x 1 = ⁻5
⁻5 x 0 = 0
⁻5 x ⁻1 = 5

Multiply each of the following.

1. ⁻6 x 7 _____

2. ⁻5 x ⁻6 _____

3. 9 x 8 _____

4. ⁻2 x 25 _____

5. 18 x 3 _____

6. ⁻4 x ⁻15 _____

7. ⁻25 x ⁻25 _____

8. 84 x ⁻6 _____

9. ⁻5 x ⁻195 _____

10. 50 x 101 _____

11. ⁻14 x ⁻2 _____

12. 15 x ⁻15 _____

To multiply more than two numbers, multiply two at a time working from the left. Be careful to write the correct sign with each product. Show all work in the following.

13. 3 x 3 x ⁻5 _____

14. ⁻4 x ⁻5 x 6 _____

15. ⁻4 x 6 x ⁻3 _____

16. ⁻10 x 10 x ⁻10 _____

17. 6 x ⁻7 x ⁻8 _____

18. ⁻8 x ⁻5 x ⁻10 _____

19. ⁻20 x 5 x 2 x ⁻10 _____

20. ⁻2 x 3 x ⁻4 x ⁻5 _____

21. 5 x 5 x 10 x 10 _____

DIVISION OF INTEGERS

Division of integers follows the same rules as multiplication.
The quotient of two positive numbers is positive: $42 \div 7 = 6$
The quotient of a positive and a negative number is negative: $42 \div {}^-7 = {}^-6$
The quotient of two negative numbers is positive: ${}^-42 \div {}^-7 = 6$

EXAMPLES **To find the solution for a combination of products and quotients, work with two numbers at a time from left to right.**

$$12 \times {}^-8 \div 16 \qquad\qquad {}^-144 \div {}^-24 \times {}^-9$$
$$^-96 \div 16 \qquad\qquad\qquad 6 \times {}^-9$$
$$^-6 \qquad\qquad\qquad\qquad {}^-54$$

Find each quotient.

1. $54 \div {}^-9$

2. $^-110 \div {}^-10$

3. $^-72 \div 9$

4. $^-200 \div {}^-25$

5. $625 \div 25$

6. $^-81 \div {}^-9$

7. $^-168 \div 8$

8. $300 \div {}^-15$

9. $^-780 \div 12$

10. $^-336 \div {}^-14$

11. $2700 \div {}^-60$

12. $^-1260 \div 30$

13. $^-1250 \div {}^-50$

14. $8000 \div {}^-20$

15. $4500 \div 5$

Find each answer by working from left to right.

16. $14 \times 5 \div {}^-7$

17. $60 \div {}^-10 \div {}^-2$

18. $^-85 \div {}^-17 \times {}^-5$

19. $^-400 \div 20 \div {}^-20$

20. $225 \times 60 \div {}^-50$

21. $15 \times {}^-10 \times {}^-2$

22. $1000 \div {}^-200 \times {}^-10$

23. $^-75 \div 5 \times 4$

24. $6 \times 6 \times {}^-6$

25. $^-18 \times 15 \div {}^-45$

26. $2000 \div 40 \div 50$

27. $10 \times 10 \times {}^-9$

28. $^-810 \div {}^-9 \div {}^-18$

29. $504 \div 9 \div {}^-8$

30. $1500 \div {}^-100 \times 20$

MULTIPLICATION AND DIVISION OF RATIONAL NUMBERS

REMEMBER: Positive and negative fractions, decimals and other rational numbers can be multiplied and divided. Do the operation as you would for positive numbers. The sign follows the rules for integers.

Positive x Positive = Positive
Positive x Negative = Negative
Negative x Negative = Positive

The same rules are true of division.

Dividing by a fraction is the same as multiplying by the inverse. $5 \div \frac{1}{4}$ is like saying "How many quarters are there in 5 dollars?"

EXAMPLES

$1.3 \times (^-6)$	$1\frac{1}{3} \times {}^-\frac{1}{2}$	$^-1.5 \div (^-0.5)$	$23 \div {}^-\frac{1}{2}$
$^-7.8$	$\frac{4}{3} \times {}^-\frac{1}{2}$	3	$23 \times {}^-2$
	$^-\frac{4}{6}$ or $^-\frac{2}{3}$		$^-46$

Find the product or quotient.

1. $\frac{1}{2} \times 3$

2. $^-\frac{1}{2} \times (^-\frac{1}{3})$

3. $^-\frac{3}{4} \times (^-\frac{1}{9})$

4. $^-3.5 \times {}^-100$

5. 15×2.5

6. $3.4 \times {}^-10$

7. 25×1.4

8. $100 \times {}^-24.5$

9. $^-11 \times {}^-3.5$

10. $1.5 \div {}^-3$

11. $^-1.5 \div {}^-3$

12. $^-150 \div 30$

13. $5.5 \div {}^-11$

14. $^-180 \div {}^-\frac{1}{2}$

15. $100 \div \frac{1}{3}$

16. $^-10 \div {}^-\frac{1}{10}$

17. $80 \div {}^-\frac{1}{5}$

18. $25 \div {}^-\frac{2}{5}$

19. $4\frac{1}{2} \div \frac{1}{2}$

20. $^-\frac{1}{2} \div {}^-\frac{1}{2}$

21. $3 \div {}^-\frac{1}{3}$

22. $\frac{1}{5} \times {}^-10$

23. $^-15 \times {}^-0.3$

24. $24 \div {}^-\frac{3}{4}$

SOLVING LINEAR EQUATIONS BY ADDING AND SUBTRACTING INTEGERS

REMEMBER: A number can be added or subtracted from both sides of an equation without changing the meaning of the equation. To solve a linear equation, you must have the unknown by itself on one side of the equation.

EXAMPLES

$$x + 14 = 20$$
$$x + 14 - 14 = 20 - 14$$
$$x = 6$$

$$x - 10 = 26$$
$$x - 10 + 10 = 26 + 10$$
$$x = 36$$

$$x + 7 = 4$$
$$x + 7 - 7 = 4 - 7$$
$$x = {}^-3$$

Solve each equation. Add or subtract as needed to find the value of the unknown. Then check your answer in the original equation. Show your work.

1. $x - 6 = 9$

2. $y + 8 = {}^-3$

3. $z + 12 = 5$

4. $a + 7 = 0$

5. $b - 6 = 10$

6. $c + 23 = 10$

7. $r - 1 = {}^-1$

8. $s + 55 = 50$

9. $t - 35 = {}^-5$

10. $x + 12 = 72$

11. $y - 67 = 101$

12. $z + 45 = 7$

13. $a - 12 = {}^-11$

14. $b + 112 = {}^-200$

15. $15 + c = 10$

16. $7 + r = 92$

17. ${}^-6 + s = {}^-27$

18. $t + 135 = 45$

19. $x - 56 = {}^-27$

20. $y + 77 = 24$

21. $z - 178 = {}^-46$

22. $a + 64 = 182$

23. $b - 49 = {}^-89$

24. $c - 36 = 59$

25. $x - 104 = {}^-68$

26. $y + 54 = 7$

27. $z - 99 = 100$

SOLVING LINEAR EQUATIONS BY MULTIPLYING AND DIVIDING INTEGERS

To solve a linear equation, you must have the unknown by itself on one side of the equation. You can multiply or divide both sides by any number or variable to get the solution. (Remember you can never divide by 0.)

EXAMPLES

$$3x = {}^-21$$
$$\frac{3x}{3} = \frac{{}^-21}{3}$$
$$x = {}^-7$$

$$\frac{x}{5} = 8$$
$$(\frac{x}{5})5 = 8(5)$$
$$x = 40$$

$$\frac{98}{x} = 14$$
$$98 = 14x$$
$$\frac{98}{14} = x$$
$$7 = x$$

$${}^-5x = 80$$
$$\frac{{}^-5x}{{}^-5} = \frac{80}{{}^-5}$$
$$x = {}^-16$$

Solve each equation for the unknown. Multiply or divide to get the unknown by itself. Then check your answer.

1. $2y = 24$

2. $4x = {}^-20$

3. $7a = {}^-63$

4. $\frac{m}{6} = {}^-4$

5. $\frac{n}{9} = 8$

6. $\frac{t}{15} = 2$

7. ${}^-4k = 32$

8. $7x = {}^-35$

9. $10y = 100$

10. $\frac{z}{{}^-8} = {}^-120$

11. $\frac{225}{c} = 25$

12. $\frac{k}{19} = 4$

13. $15a = 180$

14. ${}^-7x = 455$

15. ${}^-20y = {}^-440$

16. $\frac{a}{{}^-5} = {}^-45$

17. $\frac{585}{x} = {}^-45$

18. $\frac{21}{y} = 3$

19. $32x = {}^-832$

20. $2x = 9$

21. ${}^-4x = {}^-68$

22. $\frac{x}{15} = 15$

23. $\frac{400}{y} = 20$

24. $\frac{625}{y} = {}^-25$

25. $24a = 96$

26. $10b = {}^-1000$

27. $75c = {}^-675$

SOLVING TWO-STEP EQUATIONS WITH INTEGERS

REMEMBER: The rules that you learned in the previous two lessons can be combined to slove two-step equations. First add or subtract a number from both sides of the equation. Then multiply or divide to get the variable by itself. After you have solved the equation, check the solution in the original equation.

EXAMPLES

$3x + 5 = 23$

$3x + 5 - 5 = 23 - 5$

$3x = 18$

$\dfrac{3x}{3} = \dfrac{18}{3}$

$x = 6$

$\dfrac{y}{5} - 12 = {}^-2$

$\dfrac{y}{5} - 12 + 12 = {}^-2 + 12$

$\dfrac{y}{5} = 10$

$\dfrac{y}{5}(5) = 10(5)$

$y = 50$

$7a + 24 = 3$

$7a + 24 - 24 = 3 - 24$

$7a = {}^-21$

$\dfrac{7a}{7} = \dfrac{{}^-21}{7}$

$a = {}^-3$

Use two steps to solve each equation. Check your solution in the original equation. Do your work on another paper and write your answers here.

1. $2x + 7 = 15$

2. $\dfrac{y}{3} - 2 = 3$

3. $4x - 3 = 17$

4. $\dfrac{a}{6} + 3 = 10$

5. $9x + 40 = {}^-23$

6. $\dfrac{b}{5} - 10 = 0$

7. $2c - 35 = {}^-5$

8. $\dfrac{d}{8} - 4 = {}^-2$

9. $12z - 40 = {}^-4$

10. $9a + 15 = {}^-3$

11. $15x - 10 = 35$

12. $5t + 8 = 38$

13. $\dfrac{x}{5} + 12 = 5$

14. $100a - 75 = 225$

15. $\dfrac{y}{6} - 12 = {}^-22$

16. $4c - 11 = 17$

17. $\dfrac{k}{25} + 3 = 0$

18. $12x + 100 = {}^-20$

19. $6t - 80 = {}^-5$

20. $4x + 2 = 20$

21. $8r + 40 = 4$

22. $4x + 55 = {}^-5$

23. $\dfrac{t}{20} - 7 = {}^-2$

24. $\dfrac{72}{x} - 4 = 2$

25. $12s - 3 = {}^-33$

26. $\dfrac{30}{x} + 2 = 12$

27. $3x - 1 = 9$

28. $5y - 5 = 25$

29. $12n + 4 = 40$

30. $15x - 5 = 70$

USING EQUATIONS FOR SITUATIONS DESCRIBED IN WORDS

E X A M P L E S **First write an equation. Then solve it to find the cost.**

Three melons cost $3.45.
Let *m* be the cost of one melon.

$3m = 3.45$
$m = 1.15$

Kevin bought a car, and then sold it for $500 less. He sold it for $1900.

Let *c* be the cost of the car.
$c - 500 = 1900$
$c = 1900 + 500$
$c = 2400$

For each situation use a letter for the unknown quantity.
Then write an equation and solve it.

1. Two pounds of fish cost $14.50. Find the cost of one pound. _____

2. 2.6 pounds of meat cost $14.82. Find the cost of one pound. _____

3. Laura buys a computer for $990 and sells if for $300 less. Find the selling price. _____

4. Dwayne sells a CD player for $12 more than he paid for it. He paid $42.34. Find the selling price. _____

5. Students sold 430 tickets for the school play. The ticket price was $2.50. How much money was earned? _____

6. Amelia worked for 25 hours at $6.50 an hour. How much did she earn? _____

7. Peter worked at $7.40 an hour and earned $162.80. How many hours did he work? _____

8. The Bagel Shop sold 312 bagels and earned $280.80. What was the price for a bagel? _____

9. Drew and Marsha together earned $550. Drew earned $225.15. How much did Marsha earn? _____

10. The *Daily Gazette* paid 9 delivery boys and girls a total of $189. If they each earned the same amount, how much was each worker paid? _____

11. The temperature was 55°F at 7 A.M. What was the temperature after it rose 7°? _____

12. A shirt cost $30. If there is $1.80 sales tax, what is the cost with tax? _____

13. Groceries cost $145.86 with $4.42 in tax. What is the cost without tax? _____

14. A computer table costs $85 alone or $120 with a chair. How much does the chair add to the cost? _____

15. If 7 pounds of potatoes cost $4.13, what is the cost of one pound? _____

16. If a car travels 123 miles in 3 hours, what is its average speed? _____

17. If a plane travels 2200 miles in 4 hours, what is its average speed? _____

18. If the temperature is 67°F and 4 hours later it is 9° higher, what is the temperature 4 hours later? _____

19. If a lamp costs $17.50 without tax and $18.55 with tax, what is the amount of the tax? _____

20. If a pound of apples costs $1.45, what is the cost of 9 pounds of apples? _____

REMEMBER: An inequality identifies a set of numbers less than or greater than some given number or numbers. One number is less than another if it is to the left on the number line. Thus, 2 is less than 3 and -3 is less than 5.

The symbol < means that the number on the left is less than the number on the right, 2 < 5.

The symbol > means that the number on the left is greater than the number on the right, 5 > 2.

The symbol ≤ means "is less than or equal to." So, 2 ≤ 2 and -3 ≤ 2.

Similarly, ≥ means "is greater than or equal to." So, 2 ≥ 2 and 5 ≥ 2.

EXAMPLE **The inequality $x < 7$ has many solutions: 6, 5, 3.5, 0, -4.8 are just a few.**

The graph of the solution to the inequality $x < 7$ is shown below.
The circle at the endpoint is hollow showing that the endpoint, 7, is not included in the set.

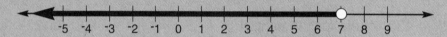

The graph of the solution to the inequality $x \geq$ -3 is shown below.
The circle at the endpoint is filled showing that the endpoint, -3, is included in the set.

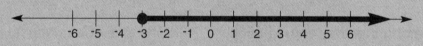

Write "true" or "false" to indicate whether each inequality is true or false.

1. 9 < 10

2. 3 ≤ -3

3. 0 > -5

4. -2 ≤ -4

5. 9 ≥ 3

6. 0 < 2

Write three numbers that are solutions for each inequality.

7. $a < 6$

8. $y \geq$ -1

9. $x < 0$

10. 3 > x

11. -1 ≤ t

12. $x <$ -2

Draw a graph on the given number line to show the solution set for each inequality.

13. $x \leq 3$

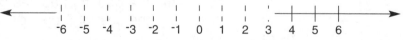

COMPOUND INEQUALITIES

REMEMBER: An inequality can give two conditions for a set of numbers. This can be done in one inequality or in two inequalities joined by the word "and" or "or."

EXAMPLE **The inequality $-3 \leq x < 5$ indicates the set of numbers that are greater than or equal to -3 and at the same time less than 5. Some solutions would be -3, 0, 2.5.**

The solution set is shown on the number line below.

Draw a graph on the number line showing the solution for each inequality.

1. $-2 < x \leq 4$

2. $-1 \leq x \leq 5$

Write the compound inequality represented by each graph.

3.
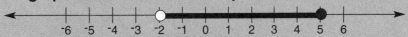
_____

4.
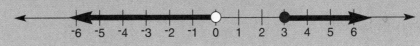
_____

REMEMBER: A compound inequality can be described using two equations. When "and" is used the solutions must satisfy both inequalities. When "or" is used the solutions must satisfy either inequality.

EXAMPLE **Draw the graph that satisfies the inequalities $x \leq 5$ and $x > -2$.**

Draw the graph that satisfies the inequalities $x < 0$ or $x \geq 3$.

Draw a graph to satisfy each set of inequalities.

5. $x > -2$ and $x < 3$

6. $x < -2$ or $4 < x$
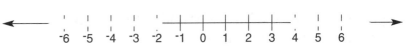

POSTTEST

1. On the number line which number is farther from 0, ⁻8 or 15? _____

2. Use arrows on the number line to add ⁻3 + 7.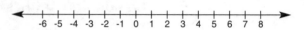

3. Add: 120 + (⁻52) _____

4. Add: (⁻10.5) + (⁻4.5) _____

5. Subtract: 28 – (⁻9) _____

6. Subtract: ⁻19 – 5 _____

Add or subtract as indicated. Show your work for each step.

7. (⁻23) + 40 + (⁻9)

   _____

8. $(-3\frac{1}{4}) + 4\frac{1}{4} + (-11\frac{1}{4})$

   _____

9. (⁻65) – (⁻23) – (⁻16)

   _____

10. ⁻7.2 + 19.3 – (⁻14.4) + 1.6

    _____

11. Find the balance in your checking account after the following transactions: Deposit $800. Write checks for $290.25, $15.45. Deposit $25.

    Balance = _____

12. Find the position of a weather balloon that starts at sea level and has the following changes: up 3600 ft; down 115 ft; up 67 ft; down 724 ft.

    Position = _____

Multiply or divide each of the following as indicated.

13. ⁻9 x 7

    _____

14. ⁻72 ÷ ⁻6

    _____

15. 8 x 1.5

    _____

16. ⁻420 ÷ ⁻7 ÷ 2

    _____

17. 12 x 7 ÷ ⁻0.3

    _____

18. $2000 ÷ ⁻100 \times 2\frac{1}{4}$

    _____

Solve each equation.

19. x – 3 = 14

    _____

20. 5x = ⁻85

    _____

21. $\frac{a}{5} = 2.4$

    _____

22. 2x – 12 = 34

    _____

23. 3y – 25 = 2

    _____

24. $\frac{z}{3} - 8 = ⁻5$

    _____

25. Pat worked at $6.40 an hour and earned $160. How many hours did he work?

    Answer: _____

CUMULATIVE REVIEW

For each figure, add to find the perimeter in inches or in centimeters.
Then convert your answer to feet or to meters.

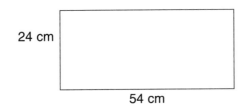

1. Perimeter = _____ centimeters

2. Perimeter = _____ meters

3. Perimeter = _____ inches

4. Perimeter = _____ feet

Perform the following operations. Carry or borrow whenever necessary.

5.   6' 3"
   +2' 6"

6.   4' 7"
   +3' 8"

7.   9' 3"
   −1' 6"

8.   3 min 20 sec
   +4 min 50 sec

9.   4 hr 16 min
   −2 hr 38 min

10.   2' 4"
    x 3

11. $\frac{1}{4}$ day = _____ hr

Circle the more likely unit for each measurement.

12. Weight of an aspirin: 1 kilogram   1 gram

13. Height of a giant redwood tree: 80 kilometers   80 meters

14. Temperature of boiling water: 100°C   100°F

Convert by multiplying or dividing.

15. 6 gallons = _____ quarts

16. 3 years = _____ months

17. 120 hours = _____ days

18. 18 feet = _____ yards

19. 48 ounces = _____ pounds

20. 16 tons = _____ pounds

Find the hours and minutes between the two given times.

21. 10:40 A.M. to 2:30 P.M.

22. 4 A.M. Mon. to 9 A.M. Wed.

Find the average (mean), median, and mode of each group of numbers.

23. 63, 68, and 67

24. Test scores: 86 (twice) and 96 (three times)

25. Draw a line graph to show the change in population in Yorktown from 1930 to 2000. Title your graph.

Year	Population	Year	Population
1930	6000	1970	13,000
1940	7000	1980	16,000
1950	11.000	1990	21,000
1960	10,000	2000	24,000

Use the spinner to find the probability of each outcome.

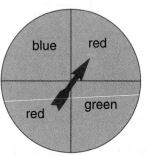

26. A red.

27. A blue, followed by a green.

Find the area of each figure.

28. $A = \frac{1}{2} bh$

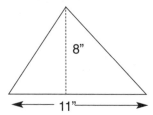

8"

11"

Area = _____

29. $A = bh$

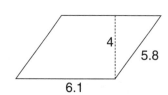

4

5.8

6.1

Area = _____

30. $1 = \pi r^2$ ($\pi = 3.14$)

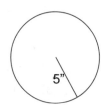

5"

Area = _____

Add or subtract.

31. $15 - 10 - (^-6)$ _____

32. $^-12 + 4.5 - (^-1.5)$ _____

Perform the indicated operations.

33. $5(^-2.5) - 6(1.4)$ _____

34. $^-3(1.5) - 7(^-2.2)$ _____

Solve each equation.

35. $x + 11 = 25$ _____

36. $a - 9.5 = ^-3.4$ _____

37. $2x + 1 = 6$ _____

38. $\frac{a}{3} - 2 = ^-1$ _____

Jim worked for 8 hours and received $58.

39. What was his hourly rate of pay?

_____

40. How much would he earn in 5 hours?

_____

TEST

CHAPTERS 1-8

NAME _____

Circle the letter of the correct answer.

SAMPLE

$7 + 35 + 10 = ?$

a. 42   **b.** 115   **c.** 52   **d.** 62   **e.** 40

Solution

$$\begin{array}{r} {}^1 7 \\ 35 \\ +10 \\ \hline 52 \end{array}$$

The sum is 52. Circle the letter **c.**

---

1. $\begin{array}{r} 115 \\ 23 \\ +240 \\ \hline \end{array}$ **a.** 378   **b.** 370   **c.** 368   **d.** 468   **e.** 470

2. $\begin{array}{r} \$32.51 \\ + \ 9.85 \\ \hline \end{array}$ **a.** $40.26   **b.** $41.26   **c.** $40.36   **d.** $41.36   **e.** $42.36

3. $\begin{array}{r} 1\frac{1}{4} \\ +3\frac{1}{4} \\ \hline \end{array}$ **a.** $4\frac{1}{4}$ **b.** $3\frac{1}{2}$ **c.** $4\frac{1}{2}$ **d.** $2\frac{1}{4}$ **e.** $4\frac{1}{16}$

4. $\begin{array}{r} 7431 \\ -2410 \\ \hline \end{array}$ **a.** 5021   **b.** 4129   **c.** $3.76   **d.** $3.66   **e.** $16.34

5. $\begin{array}{r} \$30.00 \\ - \ 26.34 \\ \hline \end{array}$ **a.** $4.76   **b.** $4.66   **c.** $3.76   **d.** $3.66   **e.** $16.34

6. $\begin{array}{r} 3\frac{1}{2} \\ -1\frac{1}{4} \\ \hline \end{array}$ **a.** $2\frac{1}{4}$ **b.** $1\frac{1}{2}$ **c.** $2\frac{1}{2}$ **d.** $4\frac{3}{4}$ **e.** 2

7. $\begin{array}{r} 72 \\ \times 35 \\ \hline \end{array}$ **a.** 1290   **b.** 2190   **c.** 2290   **d.** 576   **e.** 2520

8. $\begin{array}{r} 6.0 \\ \times 4.3 \\ \hline \end{array}$ **a.** 2.58   **b.** 25.8   **c.** 258   **d.** .258   **e.** 2520

9. $\frac{3}{4} \times \frac{1}{2} = ?$ **a.** $\frac{3}{8}$ **b.** $\frac{3}{4}$ **c.** $\frac{4}{6}$ **d.** $\frac{1}{2}$ **e.** $\frac{6}{4}$

10. $3\overline{)507}$ **a.** 102   **b.** 169   **c.** 104   **d.** 179   **e.** 12

Circle the letter of the correct answer.

11. Evaluate the expression $3x + 2$ when $x = 5$

    a. 17   **b.** 7   **c.** 13   **d.** 25   **e.** 13

12. $7\overline{)64.33}$ **a.** .919   **b.** 9.19   **c.** 91.9   **d.** 81.8   **e.** .818

13. $\frac{1}{3} \div \frac{1}{2} = ?$ **a.** $\frac{1}{5}$ **b.** $\frac{1}{6}$ **c.** $\frac{2}{3}$ **d.** $\frac{3}{2}$ **e.** $\frac{2}{5}$

14. Round 7620 to the nearest hundred.

    a. 7000   **b.** 8000   **c.** 7600   **d.** 7700   **e.** 7620

15. What would be the new price of an item marked $700 with a 20% discount?

    a. $140   **b.** $720   **c.** $680   **d.** $840   **e.** $560

16. A jacket costs $18.75. How much change should you receive from a $20 bill?

    a. $1.25   **b.** $2.25   **c.** $2.75   **d.** $0.25   **e.** $18.55

17. There are 8 school buses. Each bus contains 47 students.
    How many students are there in all?

    a. 3076   **b.** 55   **c.** 572   **d.** 376   **e.** 6

18. Find the perimeter of a rectangle with length 12 m and width 9 m.

    a. 21 m   **b.** 42 m   **c.** 30 m   **d.** 108 m   **e.** 11,664 m

19. Find the area of the rectangle described in Exercise 18.

    a. 96   **b.** 21   **c.** 42   **d.** 108   **e.** 11,664

20. Jay is 70" tall. How tall is he in feet and inches?

    a. 6'   **b.** 5'6"   **c.** 5'10"   **d.** 5'8"   **e.** 7'

21. $4^3 = ?$

    a. 12   **b.** 16   **c.** 64   **d.** 7   **e.** 81

22. The base of a triangle is 5, and the height is 3. Find the area.

    a. 12   **b.** 30   **c.** 15   **d.** 8   **e..** 7.5

23. Which number is one thousand twenty-four?

    a. 124   **b.** 10,024   **c.** 1024   **d.** 1240   **e.** 10,240

24. Find the median: 3, 6, 7, 9, 15

    a. 8   **b.** 6   **c.** 7   **d.** 7.5   **e.** 40

25. How many ounces are there in $2\frac{1}{2}$ pounds?

    a. 30   **b.** 40   **c.** 64   **d.** 6.4   **e.** 24

Circle the letter of the correct answer.

26. Reduce $\frac{6}{30}$ to lowest terms.
    a. $\frac{1}{30}$ **b.** $\frac{1}{5}$ **c.** $\frac{1}{6}$ **d.** $\frac{3}{5}$ **e.** $\frac{6}{5}$

27. Find $\frac{1}{4}$ of $40.80.
    a. $10.20   **b.** $10.80   **c.** $11.20   **d.** $11.80   **e.** $12

28. Find 15% of $30.
    a. $3.00   **b.** $4.50   **c.** $6.00   **d.** $1.80   **e.** $450.00

29. Convert $\frac{3}{5}$ to a percent.
    a. 40%   **b.** 60%   **c.** 67%   **d.** 35%   **e.** 6%

30. Convert 18% to a decimal.
    a. 1.8   **b.** .18   **c.** 18   **d.** .18%   **e.** 1800

31. A pair of shoes costs $65. The tax rate is 5%. Find the total cost with tax.
    a. $68.25   **b.** $67.25   **c.** $67.15   **d.** $65.05   **e.** $3.25

32. A garden is in the shape of a rectangle. It is 30 feet long and 20 feet wide. The owner wants to build a fence around it. How many feet of fencing should the owner buy?
    a. 600 ft   **b.** 50 ft   **c.** 60 ft   **d.** 360,000 ft^2 **e.** 100 ft

33. If a coin is tossed once, what is the probability of a heads?
    a. 1   **b.** $\frac{1}{3}$ **c.** −2   **d.** $\frac{1}{2}$ **e.** $\frac{1}{4}$

34. How many meters are in 900 centimeters?
    a. .9 m   **b.** 9 m   **c.** 90 m   **d.** 9000 m   **e.** 90,000 m

35. If a car uses 12 gallons of gas to go 348 miles, how many miles does it get per gallon?
    a. 29   **b.** 21.5   **c.** 4176   **d.** 360   **e.** 28

36. A circle has a radius of 3 centimeters. Find the area in square centimeters. Use the formula $A = \pi \times r^2$. (Use 3.14 for π.)
    a. 28.26   **b.** 18.84   **c.** 6.28   **d.** 9.42   **e.** 88.7364

37. Which angle is the right angle?
    a.   **b.**   **c.**   **d.**   **e.**

40°   60°   90°   120°   20°

Circle the letter of the correct answer.

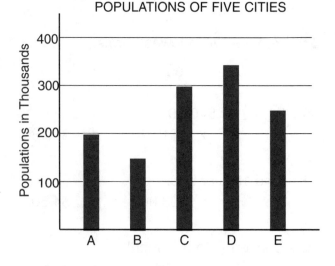

POPULATIONS OF FIVE CITIES

38. Which city has the largest population?
   a. A   **b.** B   **c.** C   **d.** D   **e.** E

39. What is the population of the smallest city?
   a. 350,000   **b.** 100,000   **c.** 150,000
   d. 200,000   **e.** 500,000

40. How many more people live in city D
   than in city A?
   a. 550,000   **b.** 100,000   **c.** 150,000
   d. 200,000   **e.** 500,000

41. Two angles of a triangle measure 90° and 30°. Find the third angle.
   a. 90°   **b.** 120°   **c.** 50°   **d.** 40°   **e.** 60°

42. Choose the largest number.
   a. .903   **b.** .90   **c.** .099   **d.** .93   **e.** .9009

43. Find the value of *n* where $\frac{6}{n} = \frac{15}{20}$
   a. 9   **b.** 8   **c.** 4   **d.** $4\frac{1}{2}$ **e.** 50

44. Find the interest on $700 for 1 year at 6%.
   a. $48   **b.** $706   **c.** $76   **d.** $24   **e.** $42

45. Add or subtract as indicated: 65 − (‾15) − 90
   a. ‾40   **b.** ‾10   **c.** 40   **d.** 10   **e.** 140

46. Find the position of a weather balloon that starts at sea level and
   has the following changes: Up 950 ft; down 75 ft; up 52 ft; down 95 ft.
   a. 878 ft   **b.** 918   **c.** 728   **d.** 832   **e.** 1152

47. Multiply or divide as indicated: 850 ÷ (‾17) x 3
   a. ‾150   **b.** 150   **c.** 836   **d.** 16.6   **e.** 10

48. Slove the equation for *x*, 3*x* − 5 = 19
   a. 4   **b.** 5   **c.** 6   **d.** 8   **e.** 9

49. Add or subtract as indicated: 11.4 − 13.7 − (‾6.5)
   a. ‾12.1   **b.** 18.6   **c.** 4.2   **d.** ‾4.2   **e..** 31.6

50. Find the balance in your checking account after the following transactions:
   Deposit $400. Write checks for $170.15, $55.45; Deposit $75
   a. $249.40   **b.** $439.70   **c.** $700   **d.** $402.70   **e.** $249.80